Climate Change and Water Resources: Challenges and Solutions towards an Imminent Water Crisis

Climate Change and Water Resources: Challenges and Solutions towards an Imminent Water Crisis

Editor: Erick West

www.callistoreference.com

Callisto Reference,
118-35 Queens Blvd., Suite 400,
Forest Hills, NY 11375, USA

Visit us on the World Wide Web at:
www.callistoreference.com

ISBN: 978-1-64116-835-9 (Hardback)

Cataloging-in-Publication Data

Climate change and water resources : challenges and solutions towards an imminent
water crisis / edited by Erick West.
    p. cm.
Includes bibliographical references and index.
ISBN 978-1-64116-835-9
1. Water-supply--Climatic factors. 2. Water-supply--Management. 3. Water resources development.
4. Climatic changes. 5. Water--Pollution. 6. Water quality management. I. West, Erick.
TD353 .C55 2023
628.1--dc23

Table of Contents

Preface

It is often said that books are a boon to mankind. They document every progress and pass on the knowledge from one generation to the other. They play a crucial role in our lives. Thus I was both excited and nervous while editing this book. I was pleased by the thought of being able to make a mark but I was also nervous to do it right because the future of students depends upon it. Hence, I took a few months to research further into the discipline, revise my knowledge and also explore some more aspects. Post this process, I begun with the editing of this book.

Water resources are sources of water, such as groundwater, rivers, springs, streams, reservoirs and lakes that are beneficial for human beings. The variability of these water resources is affected by climate change, which results in shifts in weather and temperature patterns. The rise in temperature has an impact on the hydrological cycle, as it increases the rate of evaporation of the surface water along with vegetation transpiration. The primary effects of climate change on water resources are an increase in the frequency of droughts and floods, increased temperature, decreasing snow cover and wildfires. As a result, it affects the water storage in the sub-surface and surface reservoirs. The building of resilient societies and ecosystems involves dealing with the present challenges of water crisis and climate change by adopting sustainable water management strategies. This book unfolds the impact of climate change on water resources. It consists of contributions made by international experts. This book is meant for students who are looking for an elaborate reference text on the challenges and solutions related to an imminent water crisis.

I thank my publisher with all my heart for considering me worthy of this unparalleled opportunity and for showing unwavering faith in my skills. I would also like to thank the editorial team who worked closely with me at every step and contributed immensely towards the successful completion of this book. Last but not the least, I wish to thank my friends and colleagues for their support.

Editor

Water Resource Management in Dry Zonal Paddy Cultivation in Mahaweli River Basin, Sri Lanka: An Analysis of Spatial and Temporal Climate Change Impacts and Traditional Knowledge

Sisira S. Withanachchi [1,*], **Sören Köpke** [2], **Chandana R. Withanachchi** [3], **Ruwan Pathiranage** [4] **and Angelika Ploeger** [1]

[1] Department of Organic Food Quality and Food Culture, University of Kassel, Nordbahnhofstr. 1a, 37213 Witzenhausen, Germany; E-Mail: a.ploeger@uni-kassel.de

[2] Institute for Social Sciences, Technische Universität Braunschweig, Bienroder Weg 97, D-38106 Braunschweig, Germany; E-Mail: soeren.koepke@tu-braunschweig.de

[3] Department of Archaeology and Heritage Management, Rajarata University, Mihintale 50300, Sri Lanka; E-Mail: chandanawithanachchi@gmail.com

[4] Eco-collective Research Association, Colombo 00200, Sri Lanka; E-Mail: pathiranageruwanefc@gmail.com

* Author to whom correspondence should be addressed: E-Mail: sisirawitha@uni-kassel.de;

External Editor: Monica Ionita-Scholz

Abstract: Lack of attention to spatial and temporal cross-scale dynamics and effects could be understood as one of the lacunas in scholarship on river basin management. Within the water-climate-food-energy nexus, an integrated and inclusive approach that recognizes traditional knowledge about and experiences of climate change and water resource management can provide crucial assistance in confronting problems in megaprojects and multipurpose river basin management projects. The Mahaweli Development Program (MDP), a megaproject and multipurpose river basin management project, is demonstrating substantial failures with regards to the spatial and temporal impacts of climate change and socioeconomic demands for water allocation and distribution for paddy cultivation in the dry zone area, which was one of the driving goals of the project at the initial stage. This interdisciplinary study explores how spatial and temporal climatic changes and uncertainty in weather conditions impact paddy cultivation in dry zonal areas with competing stakeholders' interest in the Mahaweli River Basin. In the framework of embedded design in the mixed methods research approach, qualitative data is the primary source while quantitative analyses are used as supportive data. The key findings from the research analysis are as follows: close and in-depth consideration of spatial and temporal changes in climate systems and paddy farmers' socioeconomic demands altered by seasonal changes are important factors. These factors should be considered in the future modification of water allocation, application of distribution technologies, and decision-making with regards to water resource management in the dry zonal paddy cultivation of Sri Lanka.

Keywords: monsoon climate; trends in precipitation; extended and short-term drought; floods; water availability; seasonal cultivation system (*yala* and *maha*); cross-scale dynamics and effects; Mahaweli Development Program (MDP)

---

1. Introduction

Integrated approaches to natural resource management have to consider diverse spatial and temporal scales [1]. River basin management directly and indirectly affects the way socioeconomic needs compete with ecological conditions. Thus, water basins are spatially and temporally entwined with natural scales (such as climate, hydrology, and ecology), socioeconomic scales (such as riparian communities, harvesting seasons, and agricultural fields), and political-administrative scales (such as localities and management zones) [2,3]. Cross-scale dynamics and effects in water resource management are a challenging proposition in the environmental policy-making process. As Termeer *et al.* [4] illustrate, how to govern and manage multi-scale problems is an open and important question in the policy-making process, where contemporary society interconnects with the complex nexus between nature and humans. Cross-scale dynamics and the effects of climate change ought to be an important consideration in the institutionalization of river basin management, especially regarding water allocation and distribution. Particularly, variations in climate and long-term weather patterns influence spatial and temporal scales in water resource management.

The dry zonal area in Sri Lanka contributes 70% of national paddy cultivation. As the report on Climate Change Vulnerability in Sri Lanka [5] states, the dry zonal area is highly vulnerable due to a prolonged drought season and diminishing precipitation. Paddy cultivation in dry zonal areas mainly depends on irrigated water. Therefore, paddy cultivation is mainly based on the river basin-oriented complex water management system. Paddy is a water-sensitive crop in every agronomic period. There are three agronomic periods in paddy cultivation: (1) vegetative (germination to panicle initiation), (2) reproductive (panicle initiation (PI) to heading), and (3) grain filling and ripening or maturation [6]. Changes in flow regimes in the mainriver, tributaries, and canals, as well as fluctuation in precipitation, can have an impact on the quantity and quality of paddy harvest.

The Mahaweli River, the longest river in Sri Lanka, flows for 335 km, covering about 16% of Sri Lanka's land area [7]; it is a key water source in the dry zone area. Based in the Mahaweli River Basin, the Mahaweli Development Program (MDP) was initiated as an empowerment program of paddy cultivation and human settlement in accordance with the agricultural system in the dry zonal area. As a second wave of river basin management, the MDP followed the prototype of Tennessee Valley Authority (TVA) (Jeroen Warner et al. [8]). It entails the construction of large-scale dams and establishes an authority-based hydraulic bureaucracy in river basin management to accelerate economic development. However, in the context of the water-climate-food-energy nexus, the MDP is demonstrating substantial failures with regards to the spatial and temporal impacts of climate change and socioeconomic demands for water allocation and distribution for paddy cultivation in dry zone areas.

The aim of this research paper is to examine how spatial and temporal climatic changes and uncertainty in weather conditions impact paddy cultivation in the dry zonal area, taking into account stakeholders' interest in the Mahaweli River Basin and how they compete. The structure of this research paper is as follows: the theoretical framework of cross-scale dynamics and effects will be in the next section. In the following section, the materials and methods of the research study will be evaluated. The fourth section will elaborate the results with a discussion of the findings. The research paper concludes with a brief summary of the main arguments and further research possibilities.

2. Theoretical Framework: Cross-Scale Inquiries of Water Resource Management

The scale as understood in this research is based on the definition of Gibson et al. [9], where a scale consists of the spatial, temporal, quantitative, or analytical dimensions that are used to measure and study any phenomenon. Cash et al. [10] advance this definition through a multi-scale understanding of environmental, geophysical, and ecological phenomena. Temporally, biophysical and social phenomena can be observed in a range of timeframes such as seasonal, electoral, budgeting, and development projects. The spatial scales correlate with levels where the socioeconomic and human-nature interaction is processed [10].

Practically, human-environment interactions are entwined across different spatial and temporal scales [10–12]. With special attention to the thematic research area "climate system", the impacts of climate change and uncertainty in weather conditions could alter and transform societal and institutional behaviors [13,14]. In particular, formal and informal decisions on agriculture and irrigation, such as cultivating periods, crop rotation, water allocation, and distribution, depend on temporal and spatial climate scales [15,16]. Furthermore, the spatial expansion of climatic zones and long-term fluctuations in climate and weather patterns including drought conditions and floods are crucial factors in designing policies of water allocation and distribution. Changes in climate conditions are likely to modify decisions and technology implementation in water resource management. Therefore, analysis of cross-scale dynamics and effects are important in the policy-making process of natural resource management [1,4,10]. As for the cross-scale dynamics, the different temporal and spatial scales have different movements and alterations in time frames and levels. Some of these dynamics impact other scales, which are known as cross-scale effects [10,17].

In hydro-social relationships, water is embedded within a physical context and on the other side, within deep sociocultural and political economic contexts [18,19]. It is constituted through varied social and material productions of scale. For instance, a naturally flowing river can be regulated through building dams, establishing new human settlements with agricultural projects, or forming canals and industrial projects [19]. The process of demarcating or choosing a scale is neither politically neutral nor entirely based on biophysical characteristics. It is a process where different stakeholders are involved in assessment by researching, reviewing, and synthesizing data as well as policy-drafting and policy implementation [20]. Thus, river basins are hydrological units where stakeholders compete and negotiate based on their conflicting interests [21,22]. Accordingly, the river basin is not a single hydrological scale, but a multiplex of several biophysical, socioeconomic and political-administrative scales [23]. The dynamics and effects of each scale on other scales are crucial in the policy-making process [3,21,22]. Cash *et al.* [10] characterized the complex interaction between human-environment systems as a synergy of scales that causes substantial complexity in dynamics. Water allocation and distribution is a pillar of water resource management and is considered a process of managing scarce resources with competitive and contested interests, with demand under pressure from other ecological constraints [15]. Spatial and temporal climatic changes and uncertainty in weather conditions is a decisive factor in water allocation and distribution. In this research study, spatial and temporal climate change and uncertainty in weather conditions are identified as the key biophysical factors in the examination of cross-scale dynamics and effects in water resource management.

3. Materials and Methods

Field work was conducted in three divisional secretariats in the Anuradhapura district (Eppawala, Thambuthegama, and Horrowpothana), as well as in Kantale in the Trincomalee district and Dambulla in the Matale district, all within the Mahaweli River Basin area (Figure 1). Though the Horrowpothana research area is not directly located within the Mahaweli irrigated regions, its water resource management is connected with the Mahaweli river basin system and management network. These divisional secretariats are located in the dry zone of Sri Lanka. The mean annual precipitation of the dry zone is less than 1200 mm [24], average annual evaporation is 1400 mm and average temperature is about 33 °C [25]. Sri Lanka is divided into three agroecological climate zones depending on annual precipitation, including dry zone. Other zones are the wet zone (high mean annual precipitation over 2000 mm) and the intermediate zone (mean annual precipitation between 2000 mm and 1200 mm), (Figure 1) [24].

The research followed the embedded design in the mixed methods research approach [26,27], as the study deals with a pragmatic and complex problem that entails both social and ecological factors. In the framework of embedded design, qualitative data was the primary source while quantitative analyses were used as supportive data. Field research from May 2012 to April 2014, semi-structured interviews, focus group discussions, and community observation were used as qualitative data-gathering methods, including primary and secondary data. We conducted interviews with members of the local communities, as well as experts and officers who held administrative and technical positions. Some of the local people interviewed are engaged in other social and economic activities. Some of them are working as community leaders or village representatives of farmers' associations (political party-oriented or not). In terms of occupation, some are working in multiple sectors (farming and fishing, or farming and livestock). In the two phases of field research, we interviewed local people (N = 45). The interviews were conducted according to a group discussions model, with semi-structured questions. Snowball sampling was used to select the interviewees.

Figure 1. Climate zones and irrigated regions (irrigation systems/zones) in the Mahaweli Development Program, and areas where field research was conducted (authors' illustration).

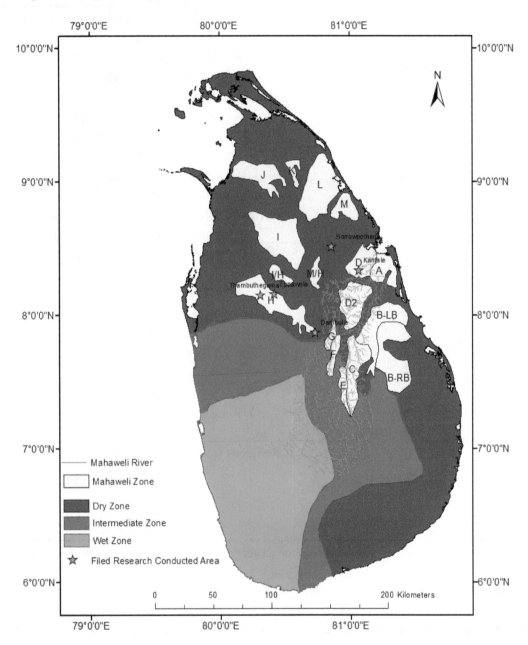

We also considered the oral history factor [28] to record the generational changes of agricultural systems in the dry zone area and the climate changes experienced in agriculture. Furthermore, experts and officials were interviewed (N = 35), from several governmental institutes, academic foundations, non-government organizations, political parties, the media, and universities (Table 1). In addition, legal documents and policy papers were other sources of information on relevant policy implementation and dynamics. The gathered data were analyzed by applying the interpretative approach within the embedded design of mixed methods research. The climate change impact and weather conditions in the spatial changes were analyzed based on recent climate research studies and databases.

Table 1. The list of institutional interviews (2012–2014).

Category of Stakeholder	Institution [1]
Agricultural administrative sector	- Mahaweli Authority, divisional branches - Divisional Agrarian Office - Divisional Irrigational Office - Water Project Field Office - Central Environmental Authority regional office - Paddy Board regional office
Non-agricultural administrative office	- Divisional Health Office - Divisional Administrative Offices - Public Health Inspector Office - Local council administrative and technical officers (secretaries, maintenance, development and welfare officers) - Village Officers (*Grama Niladari*)
Political representation	- Provincial council representatives - Local council members - Political party activists - Village-based representatives
Academics or other regional experts	- Researchers - Village-based school teachers - Regional media reporters
Non-government organization	- World Vision - RECDO (Rural Economic and Community Development Organization)
Community based group or individuals	- Religious leaders or representatives at village level - Farmers - Fishermen - Livestock sector - Retail sellers

[1.] The table indicates only the institutional or group names due to the ethical concern of the research study.

4. Results and Discussion

Spatial and temporal climate scale statuses are crucial factors in river basin management [29–31]. The decadal, annual, and seasonal variations and spatial dynamics in climate conditions are vital factors to be considered in the policy-making process of multi-purpose river basin management [11,30,31]. In the Mahaweli River Basin, spatial and temporal climatic changes influence the reconfiguration of hydro-social and hydro-political relationships in water resource management. Particularly, the flow regimes of the Mahaweli River are affected by spatial and temporal climate variation, which leads to asymmetrical opportunities in access to water resources—some social groups or stakeholders are empowered while others are disempowered or excluded. In this results and discussion section, the impact of spatial and temporal climate change and uncertainty in weather conditions are analyzed alongside socioeconomic demands and administrative and management concerns in paddy cultivation (Figure 2).

Figure 2. Combination of spatial and temporal scales in Mahaweli Basin water management.

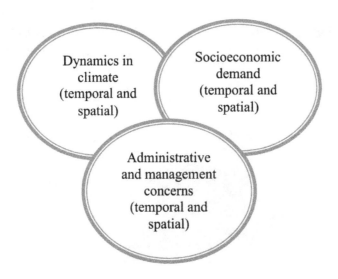

4.1. Spatial and Temporal Climate Change and Uncertainty in Weather Systems

4.1.1. Climate Scientific Observations and Predictions

Sri Lanka is mainly exposed to the Southwest monsoon (June to October), and the Northeast monsoon (December to March). Also, orographic precipitation is activated in the period from March to April and October to November because the island is located in the Inter-tropical Convergence Zone (ITCZ) [32,33]. The Mahaweli River flows from the wet and intermediate zones to the dry zone, which covers about 10,300 km^2 in the basin area [7]. The downstream area of the Mahaweli River Basin is expanding paddy cultivation. De Silva *et al.* [24] identify the main agro-ecological zones for paddy growing located in the Anuradhapura and Polonnaruwa districts, which mainly spread into the Mahaweli River Basin area.

According to Lasantha Manawadu and Nelun Fernando's [34] analysis, rainfall considerably declined in the wet zone and the intermediate zone. They demarcated the climate zones based on the isohyets of rainfall on average in the wet, intermediate, and dry zones. The 2200 mm isohyet is the line between the wet and intermediate zones and the 2000 mm isohyet is the contour between the intermediate and dry zones [34]. Though, the mean annual precipitation of the dry zone is less than 1200 mm [24], the field data demonstrates that the monthly precipitation in some areas of the dry zone could diminish to less than 50 mm. According to Manawadu and Fernando's [34] analysis, the geological area in the 2200 mm isohyet diminished from 41,582.75 km^2 to 37,000 km^2 between 1963 and 2002. The geological area in the 2000 mm isohyet has been diminished by about 9,723 km^2 between 1960 and 2000. In contrast, the dry zone has expanded from 25,323.84 km^2 to 57,227.43 km^2 (Figure 3). This is an approximate doubling of the dry zone area over the last 40 years [34]. De Silva *et al.* [24] also analyzed significant spatial changes in climate zones as a consequence of climate change. Accordingly, the spatial expansion of the dry zone would impact water allocation and distribution, and paddy cultivation. In contrast, Muththuwatta and Liyanage [35] predict that the intermediate zone would enlarge by 21.8% and the dry zone would dwindle by 8.1% by 2050 when compared to present land areas. They calculate the shifting boundaries by observing precipitation patterns from 1970. However, the important dynamic of these analyses is that climate change would alter the climate zones and increase the potential for drought conditions in the dry zone area.

Figure 3. Isohyet changes in climate zone in Sri Lanka from 1962 to 2002. (The map is illustrated based on the analysis of the Manawadu and Fernando research study [34].)

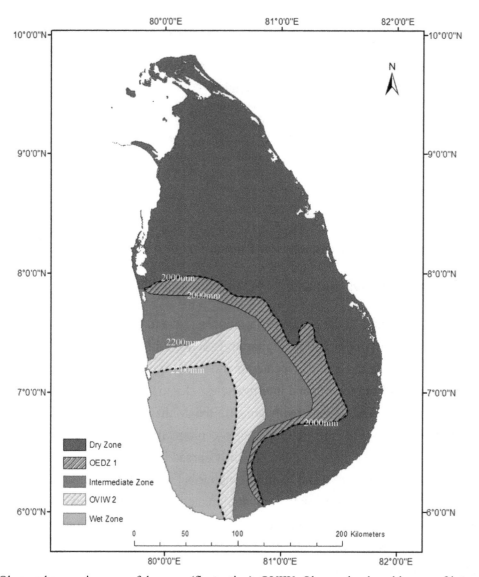

OEDZ: Observed expansion area of dry zone (fluctuating); OVIW: Observed vulnerable area of intermediate zone and wet zone (fluctuating).

Precipitation has been dramatically altered due to climate change and the uncertainty of weather patterns in the Mahaweli River Basin. The upstream areas are experiencing high fluctuations of precipitation with further declining rainfall. W.W.A Shantha and J.M.S.B Jayasundara [36] predict a 16.6% reduction in rainfall in the upper catchment areas of the Mahaweli River Basin in the highlands by 2025. Precipitation in the upper catchment area is experiencing an approximate 39.12% decline, which is a considerable reduction over the last 100 years [36]. Based on these estimations, the seasonal flow regime may drastically reduce as a general average. Moreover, Manfred Domroes and Dirk Schaefer [31] observe substantial reduction of annual average precipitation in Anuradhapura within the periods of 1895 to 1996 and 1960 to 1996 during their research study, which calculates the trends of temperature and rainfall changes in Sri Lanka. Eriyagama *et al.* [37] estimate that mean air temperatures may increase by about 0.9 °C to 4 °C by 2100 and the average rainfall will decrease. According to the IPCC report [38], the air temperature in Sri Lanka increased 0.016 °C per year. Within the period of 1990 to 2011, the minimum air temperature has increased by about 0.015 °C·y^{-1} [39]. This change in the air temperature reflects considerable changes in the climate conditions of Sri Lanka's dry zone.

According to the IPCC analysis [40], unpredictable weather changes can be expected due to abrupt climate change and climate variability. The El Niño southern oscillation (ENSO), which is identified as seasonal climate variability, is observed as seasonal climate variability in Sri Lanka. El Niño has a considerable impact on Sri Lankan precipitation and temperatures. The rainfall in the period from October to December intensifies, and further reduces in the period from January to March and July to August. Also, in the western hill slopes, including the upper catchment area in the basin, the intensification of rainfall in the period from October to December impacts stream flow, which causes inundation in downstream areas [41,42]. Furthermore, low pressure disturbances in the southwest Bay of Bengal and Southeast Arabian Sea generate tropical cyclones [43]. These tropical cyclones have a substantial impact on the Sri Lankan weather system due to heavy rainfall. For example, the dry zone areas are prone to flooding and tropical cyclones due to topographical conditions and mismanagement of flood control and irrigation systems (this will be discussed further in Section 4.4).

4.1.2. Traditional Knowledge in Climate Change and Uncertainty in Weather Systems

Local and traditional knowledge about climate and weather systems is often disregarded in national policy drafting or scientific analyses in climate studies. This tendency can be repeatedly observed in policy implementation and methodological applications in water resource management. Local people's experiences of climate history and baseline data are decisive and convenient sources in climate studies and may aid in the development of sustainable climate change mitigation and adaptation strategies [44,45]. Their long-term experience of adapting to climate change in the field and familiarity with traditional technologies, due to their livelihood, could allow them to successfully address the declining harvest and water allocation and distribution issues [46,47].

Field interviews with locals reveal their experiences of climate change and its impact on paddy cultivation and their livelihood, as well as traditional knowledge about water allocation and distribution. Table 2 codes the main responses in interviews and focus group discussions at the research locations. All interview data illustrate the presence of extended drought conditions throughout the last 20 years. Long-term droughts occurred with a high frequency in the annual period of May to November during the last 20 years, according to the interviewees' experience. Some interviewees recalled climate and weather pattern fluctuations over a period of 40 years, thus disclosing more empiric evidence of drought conditions and changes in usual seasonal cultivation systems. Particularly, farmers are recurrently impacted by short-term drought conditions, which may continue for a few weeks or less than one month. According to the interviewed farmers, short-term droughts could produce serious threats to agriculture and household water supply with unexpected water shortages. Short-term droughts may occur even during the October to February period, when people expect the annual Northeast monsoon, or in the inter-monsoon season in March. In particular, farmers' experiences demonstrate the recurrence of severe drought conditions during *yala* and *maha* seasons over the last 40 years. The official reports confirm the occurrence of drought conditions as well as floods, both seriously affecting the livelihoods of local people [5,48–50].

Table 2. The main responses (coded) in interviews and focus group discussions.

Research Location [1]	Responses (Coded)					
	Extended Long- and Short-Term Drought	Unexpected Flooding	Changing of Usual Seasonal Cultivation System	Lack of Water Supply for Agriculture	Losses of Paddy Harvest	Disregarding of Traditional Knowledge in Water Management
Eppawala	√	√	√	√	√	√
Thambuthegama	√	√	√	√	√	√
Horrowpothana	√	√	√	√	√	√
Kantale	√	-	√	√	√	√
Dambulla	√	-	√	√	√	√

[1] The research location represents villages in these divisional secretariat areas. The field data was collected as close to grassroots level as possible.

Local people encountered unexpected floods in the inter-monsoon period in March or April in Horrowpothana and Eppawala. Also, changes of rainfall during the Northeast monsoon (October to February) cause severe floods in these regions. According to the farmers' experiences, they used to have seven days of rain in February, called "*hathda wähi*". However, they observed disorder of this rainfall, either long-term rainfall or drought conditions in February. After this rainfall, most farmers usually aim to start the cultivation period knows as *yala* in the period of March–April. Therefore, they pay attention to changes in the weather system of the dry zone area because of economic relevance to their families. An important finding of reviewing the local-based traditional knowledge about climate and weather systems is that local people are aware of short-term weather changes and long-term changes in the climate system of the Mahaweli River Basin areas. This knowledge is either used to adapt to long-term climate change or communicated to village-level or regional officers in the Mahaweli Authority.

4.2. Water-Climate-Energy-Food Nexus in Competing Interests of Scale (Re) Configuration

The Master Plan for the MDP was assessed in 1965–1968 by the United Nations Development Program (UNDP), the Food and Agriculture Organization (FAO) team, and Sri Lankan engineers [50,51]. The program was implemented in 1970 and then restarted in 1977 under the new government. Hydropower generation and agricultural development could be observed as the main political objectives that motivated the MDP. Water allocation and distribution are based on the competitive interests of stakeholders in the MDP. The demand for water resources is shared among macro and micro hydro-electrical plants, vegetable farmers upstream (mainly in the highland of Sri Lanka), paddy farmers downstream, and industries. Through an analysis of the temporal locus of the development paradigm, it could be observed that the locus and focus of the MDP have been changing. As a result of state-oriented development programs after independence in 1947, large-scale and multi-purpose water resource management programs were politically manifested. The state-led development before the 1980s was mainly motivated by promoting welfare programs and land reform for peasant farmers, allocating lands for agriculture in semi-arid dry zone areas with relatively low population density, and promoting paddy cultivation in the national economy [52,53].

To promote paddy cultivation in the national economy, new paddy fields were constructed in Mahaweli Basin irrigation regions and then were distributed among new settlers. The downstream area of the Mahaweli River Basin has been demarcated into 13 irrigation regions (Figure 1). Each region has been divided into sub-regions from 1000 hectares to 1500 hectares in land area [54]. Furthermore, these sub-regions have been partitioned into hamlets (village units) of 150 families [55]. The aim of the divisions is a manageable irrigation region and functional human settlement system [54,55]. According to new information in 2012, there are 365,000 cultivated hectares in the Mahaweli River Basin area, including 10,049 km of canal networks [48]. Currently, approximately 2.8 million people live in the Mahaweli River basin, which is about 15% of the population of Sri Lanka [7]. There are about 166,269 households in the Mahaweli settlement areas [50]. Most settlers migrated from the central parts or western hill slopes of Sri Lanka as landless peasants [56]. They settled under the government program in the Mahaweli irrigated regions (Mahaweli Irrigation Systems/Zones) from A to M (Figure 1). Each household was permitted 1 hectare of irrigated lowland for paddy cultivation and 0.2 hectares of rain-fed highland farm steading [56]. The spatial responsibilities of the Mahaweli Authority extend beyond the physical context of the Mahaweli River Basin. Chiefly, the Udawala project in the Walawe River Basin in southern Sri Lanka and some irrigational regions including H, I, L, and M are located beyond the physical context of the Mahaweli River Basin (Figure 1). Furthermore, the irrigation regions M and M/H are located in the Yan Oya basin. These sub-river basins are connected by water diversion from the Mahaweli River. As discussed, water diversion between sub-river basins is another purpose of the MDP.

In contrast, neoliberal political transformation was the driving force of state policies after the 1980s. Within this context, the government acts as a facilitator to construct infrastructure facilities for the private sector in the context of a market economy [53]. This created vital changes in the focus of the MDP by facilitating electricity generation for the Free Trade Zones. From the 1980s onwards, the priority of the MDP shifted to establishing hydropower plants by constructing large dams upstream and in the middle reach of the Mahaweli River [52,57]. Large dam constructions are the expression of a modernization path following a Western technologic-economic rationality. Large, multi-purpose dams are the center of the hegemonic hydraulic paradigm [58], and their planning and implementation have often disregarded concerns for social and environmental sustainability. In the case of the MDP, major foreign investments have been allocated to constructing water reservoirs as the basis of hydropower generation, such as the Kotmale, Victoria Randenigala, Rantembe, Ulhitiya/Ratkinda Madu Oya Maduru Oya, Bowetenna, Udawalawe, Ukuwela, and Polgolla hydropower plants. Meanwhile, the MDP is being expanded to new spatial areas beyond the physical context of the Mahaweli River Basin, to include such ventures as the Moragahakanda and Kaluganga Rivers Development Projects, the Rambaken Oya Development Project, and the Kiwul Oya Reservoir Development Projects [50]. Under the Moragahakanda-Kaluganga Project in 2007, the government expects to enhance hydropower generation as well as drinking water purposes and cropping intensity in the Mahaweli River Basin settlement areas [50]. In 2005, 30% of national hydropower production was from the hydropower plants within the Mahaweli system [36].By contrast, 49% of national hydropower was generated under the MDP in 2011. The total power generation to the national grid from the Mahaweli system is 1975 Gigawatt hours (GWh), which is 49% of the total hydro power in Sri Lanka [50].

The gradual socioeconomic transformation from an agriculture-oriented society into a more industrial-oriented society [37] may place high demands on water resources in the MDP. The statistical analysis of the sectoral water withdrawal in Sri Lanka demonstrates the increase of industrial water use and domestic water use, and the gradual decrease of agricultural water demand. As Figure 4 shows, the irrigational usage of water has gradually declined. It is estimated that irrigational water usage in 2025 will be between 70% and 75%. However, other competing sectoral water demands challenge the demand for water in paddy cultivation. According to this trend, water usage for domestic purposes and industries are expected to gradually increase. FAO-AQUASTAT data on renewable internal freshwater resources per capita per year (actual) in Sri Lanka indicates 2482 m^3 in 2014 [59], which shifts the country into the vulnerable category (based on UN WWDR and UN-Water category thresholds [60]) compared to 2012 data. Under the conditions of high-demand water consumption and a challenging climate, the water-climate-energy-food nexus is being contested by the changing economic interests and political manifestations of natural resource management at the national level. At this juncture, MDP can be observed as a core realm of competing interests over water resources integrating local socio-economic demands at the regional level.

Figure 4. Sectoral Water Demand in Sri Lanka from 1990 to 2025. (Source: The Annual Report—Economic perspective Sri Lanka [39]).

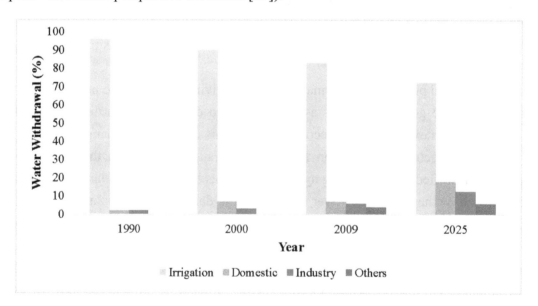

4.3. Impacts of Seasonal and Temporal Climate Change on Paddy Cultivation

Temporal and spatial changes in climate conditions and weather systems are crucial factors to take into consideration in the dry zonal water resource management. Releasing water for paddy cultivation depends on the flow regimes of the Mahaweli River and its tributaries. There are two trends reflecting seasonal flow regimes, including the impact of high precipitation and the gradually declining precipitation in catchment areas. As a result of spatial and temporal variations in climate conditions, specifically declining precipitation, rising air and soil temperature, and increasing evaporation rates in the irrigational management areas of the Mahaweli River Basin, alteration of seasonal-oriented agricultural decisions in the field study areas could be observed.

There are two main seasons of paddy cultivation in the dry zone, *maha* (October to February/March) and *yala* (March/April to July/August) which are defined on the basis of periodic precipitation. Based on the field data, the average precipitation in the *maha* season could be ranged 750 mm to 1000 mm, and average precipitation in the *yala* season could be around 500 mm or less. In the dry zone, the *maha* season is the main harvesting season. Some farmers cultivate other crops (grains or vegetables) during the *yala* and *maha* seasons, called intermediate cultivation. However, the decision to cultivate an intermediate season depends on water availability. Though agriculture in the dry zone is mainly based on irrigated water, some farmers are accustomed to using rain-fed water in the *maha* season and irrigated water in the *yala* season based on seasonal variation of precipitation. Water allocation and distribution for paddy cultivation under the MDP also operate based on this general assumption. However, this seasonal-oriented paddy cultivation is challenged by the impacts of long-term climate change and uncertainty in weather systems. Eriyagama *et al.* [37] estimate that the water requirement for paddy cultivation in the *maha* season would be 13%–23% in 2050, as compared to the period from 1961–1990 with reference to quantity and spatial distribution of precipitation and changes in mean temperature.

Figure 5. Losses to agricultural crops in hectares due to droughts and floods, 1974–2007. (Source: Sri Lanka National Report on Disaster Risk, Poverty, and Human Development Relationship, the Disaster Management Centre (DMC) of Sri Lanka [49]).

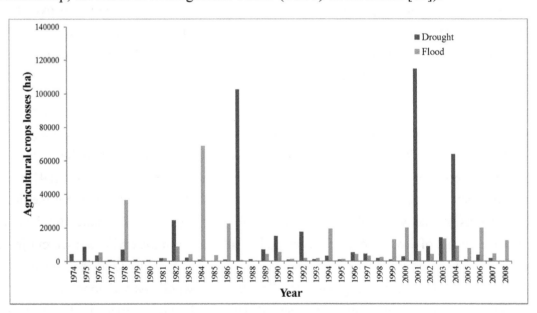

The frequent occurrence of dry weather conditions in recent years has resulted in declining overall agricultural output in downstream areas [61–63]. Particularly, paddy cultivation in the dry zonal area exposes it to short-term and extended severe drought conditions. Paddy yields at early stages of the agronomic period have been affected due to short-term droughts. Furthermore, the lack of water distribution in the reproductive phase of rice could reduce the harvest of paddy. Because of that, farmers are unable to achieve off-season harvesting and intermediate cultivation due to declining rainfall in the *yala* season. According to data from the Disaster Management Centre of Sri Lanka (DMC) [49], severe drought conditions brought about damage to agricultural land, particularly in the dry zone area. In the period from 2001 to 2004, the dry zone area was exposed to extended drought conditions in both the *yala* and *maha* periods. In the range of the research area, Hambanthota and Kurunegala are the other districts that were affected by these drought conditions. Figure 5 illustrates the losses of agricultural crops in hectares caused by droughts within the period of 1974 to 2007 [49]. The size of fields under paddy cultivation has been reduced to 66,194 hectares in the dry and intermediate zones of the Mahaweli River Basin as a result of severe drought conditions in the *yala* season in 2012 [48]. Based on the field data, the dry zone is again prone to be exposed to extended drought conditions from *maha* season in 2013 to *yala* season in 2014. These drought conditions impact paddy cultivation as well as the cultivation of other crops. The exact losses to agriculture caused by drought conditions have not yet been officially calculated. Based on the Sri Lanka Disaster Management Center's data from 1974 to 2008 on the frequency and scale of disasters, the Sri Lankan Ministry of Environment analyzes the spatial distribution of extended drought conditions that create vulnerability in the irrigation sector [5] (Figure 6). Based on this GIS-based analysis at the divisional secretariat (DS) level, the high level of vulnerability in the irrigation sector can be scrutinized in the Mahaweli D2, H, I/H, M/H, I, J, L, and M regions.

The interview data explains that paddy cultivation especially in the *yala* season was either abandoned or earned a limited harvest throughout the last 10-year period. According to the farmers' experiences, there is latent competition over water allocation during periods of limited water availability. According to the authorities of the MDP, water supply for paddy cultivation is frequently controlled and limited due to the inadequate water availability in water tanks and tributaries that are located in the Mahaweli River basin [64]. In interviews, paddy farmers repeatedly expressed deep concern over the lack of adequate water resources, especially during the *yala* season. The latest data demonstrate that in the *yala* season of 2014, about 35,000 acres in the Polonnaruwa District (MDP regions) could not be cultivated because of drought conditions [64].

Most of Mahaweli irrigation regions are affected by drought, and the situation has a reciprocal influence on current water resource management practices. The recent drought conditions are extending to Mahaweli B, C, H, and G regions including the Polonnaruwa, Girandurukotte, Galewela, Dambulla, Dehiattakandiya, Anuradhapura, Elahera, Tambuettegama, and Kandalama divisional secretariats (DS) [64]. At the field level, the spatial expansion of dry zonal conditions could be observed towards the intermediate climate zone, which points to the vulnerability of farmers in Kurunegala and Matale districts. Figure 7 demonstrates the impact of drought conditions on paddy cultivation at the divisional secretariat level [5]. According to this GIS-based explanation, a high level of vulnerability in the paddy sector can be observed in the Mahaweli H, I/H, M/H, I, and J regions, as well as the Horrowpothana divisional secretariat. Farmers reveal that they have encountered extensive losses of their paddy harvest over the last 10 years due to increased air temperature and extended drought periods in the *maha* and *yala* harvesting periods. Consequently, these effects induce an increase in surface temperature [33,65], which causes deterioration of harvest in paddy cultivation [66]. According to Matthews *et al.* [67], the increase of seasonal average temperature reduces the paddy harvest. Welch *et al.* [68] state that a higher minimum temperature decreases the paddy yield, while a higher maximum temperature could increase paddy cultivation. The evaporation of water from rivers and water tanks, diminishing soil moisture, and declining groundwater recharge affect paddy cultivation under the drought conditions [5].

Figure 6. Irrigation sector vulnerability to drought exposure.

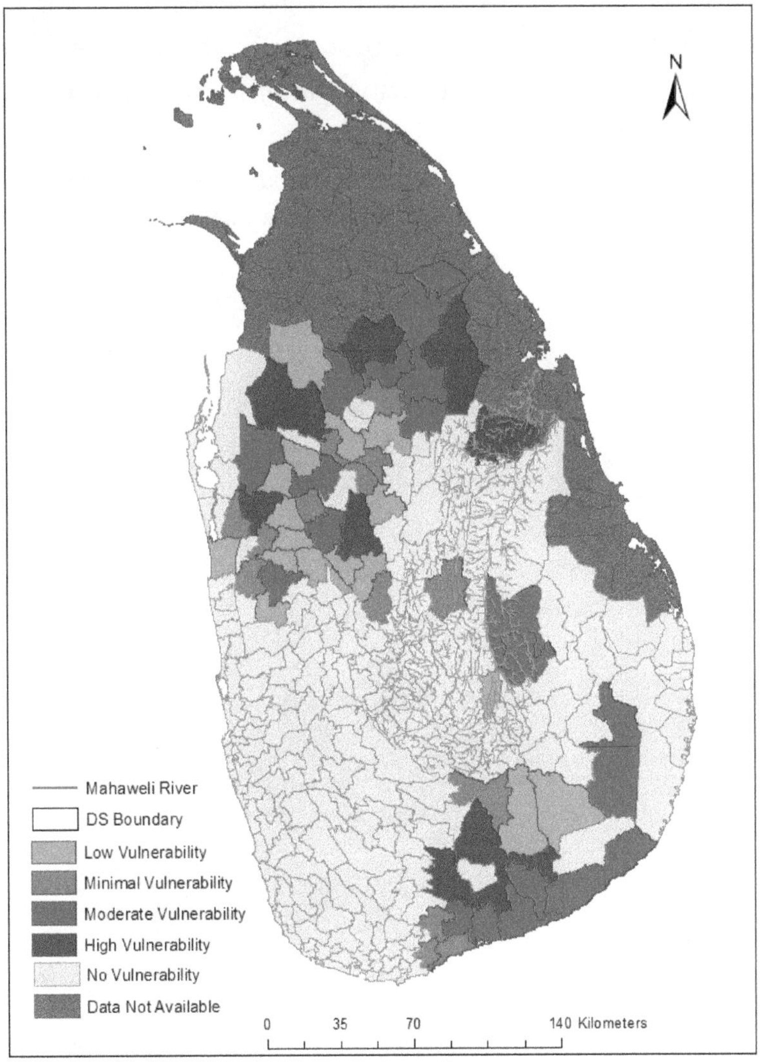

DS: Divisional Secretariat. (Map is re-illustrated based on the data from the Climate Change Vulnerability Data Book [5].)

Constant stream flow cannot be expected throughout the year in the Mahaweli River. Heavy rainfall in the upper catchments, activation of inter-monsoon rainfall in the dry zone area, and cyclone condition activation in the Bay of Bengal generate high stream flow in the river, tributaries, and water channels. These temporal changes in climate conditions and weather systems impact the seasonal orientation of paddy cultivation. Observation demonstrates that farmers are highly vulnerable to unexpected flooding in the *yala* and *maha* seasons. Their anxiety over crop protection from flooding of the farm fields is a major issue in agriculture. For example, the activation of the Northeast monsoon in the month of December 2010 severely damaged small, medium, and even large-scale irrigation projects. The DMC data [49] shows considerable damage to agricultural lands because of flood conditions from 1974 to 2008. In the dry zonal area, the Polonnaruwa, Batticaloa, Killinochchi, and Ampara districts encountered severe flood conditions over the last 34 years. As illustrated in Figure 5, substantial losses to agricultural lands (paddy and other crops) were documented in 1978, 1984, 1986, and the period from 1999 to 2008. Around 303,957 hectares of paddy were critically damaged due to drought conditions between 1974 and 2007 [69]. As the Central Bank report of Sri Lanka [63] states, the canal network and 67,900 hectares of agricultural land were submerged and 6285 hectares were destroyed. Most of these irrigated projects and agricultural lands are located in the A, B, and D irrigation regions of the MDP [63]. Due to the uncertainty in weather conditions and climate change scenarios, the seasonal dynamics of flow regimes have to be considered with regards to water productivity in agriculture. Seasonal dynamics of river flow regimes should be carefully assessed and studied, because variation of water quantity in rivers and tributaries impacts hydropower generation, irrigation, flood risk management, and water allocation, which are all components of a holistic river basin-oriented water management plan.

Figure 7. Paddy sector vulnerability to drought exposure.

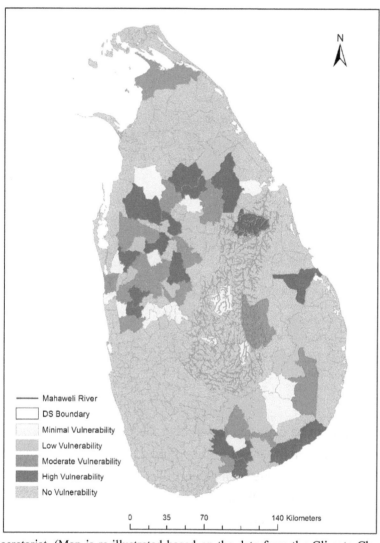

DS: Divisional Secretariat. (Map is re-illustrated based on the data from the Climate Change Vulnerability Data Book [5].)

Drought and floods cause depleted yields or harvest failure, which leads to socioeconomic hardship for small-scale farmers. As recorded from 1974 to 2007 (Figure 8), major hazards that adversely affected paddy cultivation are drought conditions (around 50.83%) and flood conditions (around 45.83%). In total, 578,014 hectares under paddy cultivation have been severely damaged due to flood and drought in the 1974–2007 period [69]. Farmers were concerned not only about limited water availability, but also about local, small-scale irrigation technologies, which are in a state of disrepair. With regards to the role of MDP authorities and other state agencies, farmers complained about the irrigation bureaucracy's failure to adequately meet their preferences in water allocation, and its lack of attention on maintenance and operation of water supply systems. Therefore, spatial and temporal impacts of climate change and climate variability with drought and flood conditions could reshuffle or alter seasonal decisions on agricultural practices.

Figure 8. Effect of hazards on paddy cultivation, 1974–2007.

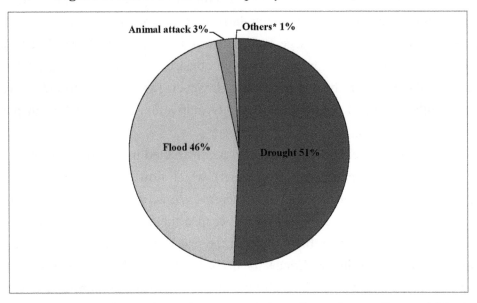

* Others: cyclone, gale, landslide, plague, frost, tsunami, forest fire, and storm (Source: Sri Lanka Disaster Knowledge Network [69]).

4.4. Modern and Traditional Application in Irrigational Water Management

Historical evidence demonstrates that water resources and riparian areas of the Mahaweli River had been used for water diversion through canalling and construction of the cascade system for agriculture and domestic consumption [70–72]. The traditional irrigation system, which was based on the strong hydraulic civilization of about 2000 years ago (around 250 B.C. to around 1100 A.D.) in the dry zonal area, is considered an appropriate system for dealing with harsh, changing climate conditions and has the capacity of adapting to long-term changes in climate [72,73]. However, a discrepancy between modern application in the MDP and traditional knowledge of irrigational water management has been observed.

With the development of the agriculture and irrigation sector in the MDP, a considerable number of village tanks (*kotu wewa*), which were common in the dry zone, were converted into paddy fields. These new paddy fields were distributed among new settlers. The purpose of the village tanks was to supply water for agriculture and other human purposes during extended dry seasons. These small tanks were used when the water level decreased in other, major tanks. The village tank (*kotu wewa*) system is considered the oldest tank system in Sri Lanka [73]. In ancient times settlements in the dry zone were located around a tank (*wewa*) rather than along or near a water channel in the dry zone area. The village tanks (*kotu wewa*) were also an essential unit of the traditional river-basin-oriented complex water management system. The tanks were stored with rainwater or the diverted water from a channel barricaded by small anicut (dams). C. M Madduma Bandara conceptualizes this system as a cascade system which is defined as a "connected series of tanks organized within a micro-catchment of the dry zone landscape, storing, conveying and utilizing water from ephemeral rivulets" ([71], p14). A cascade of tanks is constructed by 4 to 10 individual small tanks, with each tank having its own micro-catchment. All village tanks were situated within a single meso-catchment basin that was organized in extent from 15.5 km^2 to 25.8 km^2, with a model value of 20.7 km^2 in the dry zone area. The outflow from a tank was stored by a downstream tank. This water storage could be applied within the command area of the second tank. Thus, the channel runoff was continuously recycled and refilled. This system could assist to surmount the problems of irregularly distributed rainfall, non-availability of large catchment areas, and the difficulty of large tank construction [73]. Local people and archeologists assume that these village-based small tanks have been constructed based on long-term experience of climate conditions and changes in the dry zone area. However, few village tanks (*kotu wewa*) are still operated for water allocation and distribution in paddy cultivation and other human purposes at the village level in the dry zone area.

Some of the old water channel systems were also reconstructed in concrete under the MDP. In the Eppawal division, *Yoda Ela* (a water diversion channel), which flowed by nurturing the ecosystem and recharging the groundwater, is reconstructed with concrete as *Nawa Jaya Ganga*. However, the expected goals from this reconstruction cannot be achieved because the water channels are inundated in the inter-monsoon period [74]. A considerable portion of the paddy harvest is destroyed every year due to unexpected flooding conditions in the Eppawala division. Mishandling or ignorance of traditional technology in water distribution also leads to diminishing soil moisture and impacts on the paddy fields and water springs as a result of lack of recharge of groundwater.

The complex irrigational management system in the MDP weakens voluntary engagement of farmers in water resource management (Figure 9). This complexity leads to the strengthening of bureaucrats in irrigation and agricultural management [75–77]. Due to this hydrocracy [78] in water resource management in the MDP, societal demands and ecological changes are disregarded under bureaucratic control of the technical infrastructure. The highly centralized administrative and management system in the MDP hinders voluntary farmer involvement [76,77,79]. Water allocation and distribution is mainly administered by the officers in the regional office of the Mahaweli Authority in each irrigation region. The research data demonstrate that the "*kanna rasweema*" (the meeting at the cultivation period) in most regions in the Kantale and Horrowpothana divisions do not empower farmers' voluntary decision-making. At the village level, "*kanna rasweema*" is a participatory body of farmers competent in decision-making regarding water allocation and distribution. Though farmer associations are responsible for the operation and maintenance of small and minor irrigation schemes, officers in the regional office of the Mahaweli Authority supervise and operate water management. The miscommunication or ignorance of farmers' water demands is the main reason for the loss of expected profit from paddy cultivation prior to other factors, such as pricing issues or lack of storage.

The lack of participation of the farmer associations in water allocation and distribution currently leads to the malfunction of the irrigation canal system at the micro level. The spatial changes in the dry zone are not identified in order to reorganize the water allocation system. The farmers' experience and traditional knowledge are rarely acknowledged or taken into consideration within the water resource management planning of the MDP. Within the technocratic irrigational system of the MDP, farmers are either discouraged or prevented from accessing traditional applications in water resource management such as *diya bäduma*. The term *diya bäduma* refers to the gradual water allocation system from main water flow (the river) by barricading small anicut (dams). The diverted water flows into the small canal, then into a water channel in a farm field. In the *diya bäduma* system, the direct engagement of villagers is a crucial factor because the amount of water diversion, time period, and area of spreading water are decided upon the voluntary participation and will of the people. Traditional irrigational practices are considered as part of the traditional body of knowledge adapted to changing climate conditions, ecosystem, and social demands. Lack of opportunities for voluntary engagement in water resource management, especially canal management and water allocation and distribution, contribute to the malfunction of the irrigation canal system at the micro level.

Figure 9. River basin-oriented management system and political-administrative management system in the Mahaweli River Basin area (authors' analysis).

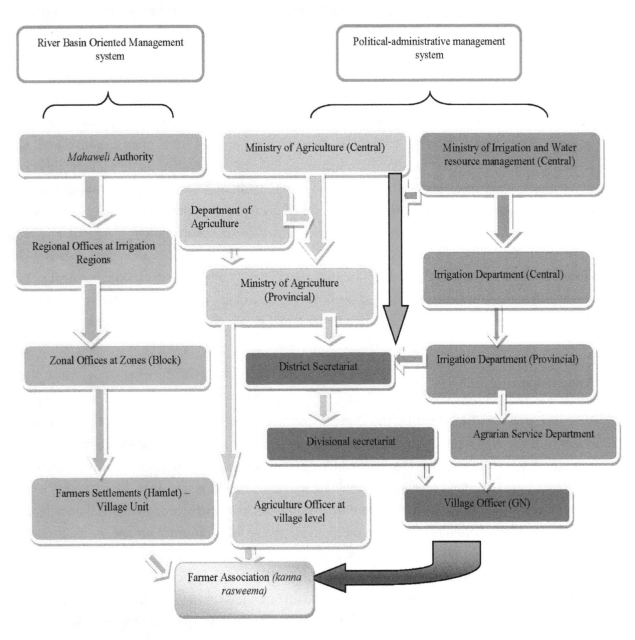

(Arrows indicate the flow of responsibilities and administrative power)

5. Conclusions

A close observation of cross-scale dynamics and effects in water resource management is crucial to taking a sustainable path to empowering rural agricultural systems. In the context of the water-food-climate-energy nexus, small-scale agriculture faces serious challenges not only from climate change and uncertainties in weather conditions, but also from socioeconomic and administrative factors. Paddy cultivation is the main source of income and the basis of food security in small-scale farming in Sri Lanka. A lack of understanding with regards to the spatial and temporal dynamics in climate, socioeconomic demand, and administrative or management functionalities could generate negative consequences for farmers in the dry zone of Sri Lanka. The challenges mentioned above should be addressed by shifting water governance policies. The existing hydrocracy in the MDP should prioritize inclusive and integrated water resource management in order to democratize the water allocation and distribution process in the irrigation sector. Furthermore, research demonstrates that water allocation and distribution in the MDP, taking into consideration competing stakeholders' interests, depend on the effects and dynamics of spatial and temporal climate change and uncertainty in weather conditions. Spatial changes in the dry zone area, taking into account the impacts of climate change, call for a substantial alteration of adaptation policies in water resource management in the MDP. Temporal changes in the climate system and seasonal-oriented agricultural decisions are affected by extended and short-term droughts and flood conditions in the dry zone area. As the main negative consequence, deteriorating paddy harvests translate into socioeconomic hardships that create vulnerability for local paddy farmers. Moreover, the MDP has to acknowledge traditional knowledge in climate and irrigation water management and ancient technological applications in the dry zone area, which has strong roots in historical hydraulic civilization. Therefore, close and in-depth consideration of spatial and temporal changes in climate systems and paddy farmers' socioeconomic demands as altered by seasonal changes are important factors. As the key findings of this research study, these factors should be considered in the future modification of water allocation and application of distribution technologies and in decision-making with regards to water resource management in dry zonal paddy cultivation in Sri Lanka.

Acknowledgments

This research was facilitated by the Department of Organic Food Quality and Food Culture at the University of Kassel, Germany and the Department of Archeology and Heritage Management at Rajarata University, Sri Lanka. Also, the DAAD-financed EXCEED program on sustainable water management coordinated by TU Braunschweig, Germany partly funded field research in May 2013. The authors gratefully acknowledge D.M.S.L.B Dissanayake and Thilanka Siriwardhana at Rajarata University, Sri Lanka for their enormous support in the field study and illustrations. Our special gratitude goes to Damien Frettsome, Amanda W. Schimunek, and Garret Field for language editing. Also, we would like to thank J. Herden, S. Sapkota, M.T Semiromi, Y. Fasihi, and Pavithra Tantrigoda for their extensive support in this research. We extend our gratitude to all interviewees in the field who dedicated their time and allowed us to participate in their meetings. Without their unconditional contributions, it would have been impossible to complete this field research study. We highly appreciate the opportunity to present our research study at the International Food System Symposium in October 2013, organized by Yale University, and critical comments from the audience. Last but not least, we would like to thank the anonymous reviewers of the Journal *Climate* for their critical and constructive comments.

Author Contributions

All authors contributed equally to this research work.

References

1. Pahl-Wostl, C. The implications of complexity for integrated resources management. *Environ. Modell. Softw.* **2007**, *22*, 561–569.
2. Lebel, L. The politics of scale in environmental assessment. In *Bridging Scales and Knowledge Systems: Concepts and Applications in Ecosystem Assessment*; Reid, W.V., Berkes, F., Wilbanks, T.J., Capistrano, D., Eds.; Island Press: New York, NY, USA, 2006; pp. 37–57.
3. Loucks, D.P.; van Beek, E.; Stedinger, J.R.; Dijkman, J.P.M.; Villars, M.T. Water resources planning and management: An overview. In *Water Resources System Planning and Management: An Introduction to Methods, Models and Applications*; UNESCO: Paris, France, 2005; pp. 3–36.
4. Termeer, C.J.A.M.; Dewulf, A.; Van Lieshout, M. Disentangling scale approaches in governance research: Comparing monocentric, multilevel, and adaptive governance. *Ecol. Soc.* **2010**, *15*, 29.
5. Ministry of Environment-Sri Lanka. *Climate Change Vulnerability Data Book*; Ministry of Environment: Colombo, Sri Lanka, 2011.
6. Moldenhauer, K.; Slaton, N. Rice growth and development. In *Rice Production Handbook;* Slaton, N.A., Ford, L.B., Bernhardt, J.L., Cartwright, R.D., Gardisser, D., Gibbons, J., Huitink, G., Koen, B., Lee, F.N., Miller, D.M. *et al,* Eds. Cooperative Extension Service, Division of Agriculture, University of Kansas: Lawrence, KS, USA, 2004; pp. 7–14.
7. Elakanda, S. Resource-based development: Experience from Mahaweli. In *CRBOM Small Publications Series No. 29*; Centre for River Basin Organizations and Management: Central Java, Indonesia, 2010; pp. 1–9.
8. Warner, J.; Wester, P.; Bolding, A. Going with the flow: River basins as the natural units for water management? *Water Policy* **2008**, *10*, 121–138.
9. Gibson, C.C.; Ostrom, E.; Ahn, T.K. The concept of scale and the human dimensions of global change: A survey. *Ecol. Econ.* **2000**, *32*, 217–239.
10. Cash, D.W.; Adger, W.F.; Berkes, P.; Garden, L.; Lebel, P.; Olsson, L.; Pritchard, O.Y. Scale and cross-scale dynamics: Governance and information in a multilevel world. *Ecol. Soc.* **2006**, *11*, 8.
11. Pahl-Wostl, C. Transitions towards adaptive management of water facing climate and global change. *Water Recour. Manag.* **2007**, *21*, 49–62.
12. Kok, K.; Veldkamp, T. Scale and governance: Conceptual considerations and practical implications. *Ecol. Soc.* **2011**, *16*, 29–45.
13. Adger, W.N.; Arnell, N.W.; Tompkins, E.L. Successful adaptation to climate change across scales. *Glob. Environ. Change* **2005**, *15*, 77–86.

14. WWAP (World Water Assessment Programme). *The United Nations World Water Development Report 4: Managing Water under Uncertainty and Risk*; UNESCO: Paris, France, 2012.

15. Molle, F.; Wester, P.; Hirsch, P.; Jensen, J.R.; Murray-Rust, H.; Paranjpye, V.; Pollard, S.; Van der Zaag, P. River basin development and management. In *Water for Food and Water for Life*; Molden, D., Ed.; International Water Management Institute (IWMI): Colombo, Sri Lanka, 2007; pp. 585–624.

16. Van Lieshout, M.; Dewulf, A.; Aarts, N.; Termeer, C. Do scale frames matter? Scale frame mismatches in the decision making process about a "mega farm" in a small Dutch village. *Ecol. Soc.* **2011**, *16*, 38–53.

17. Leibold, M.A.; Holyoak, M.; Mouquet, N.; Amarasekare, P.; Chase, J.M.; Hoopes, M.F.; Gonzalez, A. The metacommunity concept: A framework for multi-scale community ecology. *Ecol. Lett.* **2004**, *7*, 601–613.

18. Swyngedouw, E. *Place, Nature and the Question of Scale: Interrogating the Production of Nature*; Discussion Paper 5; Brandenburgische Akademie der Wissenschaften: Berlin, Germany, 2010; pp. 5–29.

19. Swyngedouw, E. The political economy and political ecology of the hydro-social cycle. *J. Contemp. Water Res. Educ.* **2009**, *142*, 56–60.

20. Lebel, L.; Garden, P.; Imamura, M. The politics of scale, position, and place in the governance of water resources in the Mekong region. *Ecol. Soc.* **2005**, *10*, 32–58.

21. Dore, J.; Lebel, L. Deliberation and scale in Mekong Region water governance. *Environ. Manag.* **2010**, *46*, 60–80.

22. Moss, T.; Newig, J. Multilevel water governance and problems of scale: Setting the stage for a broader debate. *Environ. Manag.* **2010**, *46*, doi:10.1007/s00267-010-9531-1.

23. Withanachchi, S.S.; Houdret, A.; Nergui, S.; Gonzalez, E.E.I.; Tsogtbayar, A.; Ploeger, A. *(Re)Configuration of Water Resources Management in Mongolia: A Critical Geopolitical Analysis*; International Center for Development and Decent Work (ICDD): Kassel, Germany, 2014.

24. De Silva, C.S.; Weatherhead, E.K.; Knox, J.W.; Rodriguez-Diaz, J.A. Predicting the impacts of climate change—A case study of paddy irrigation water requirements in Sri Lanka. *Agric. Water Manag.* **2007**, *93*, 19–29.

25. Jayawardanaa, D.T.; Pitawalab, H. M. T. G. A.; Ishigaa, H. Groundwater quality in different climatic zones of Sri Lanka: Focus on the occurrence of fluoride. *Int. J. Environ. Sci. Dev.* **2010**, *1*, 244–250.

26. Creswell, J.W.; Clark, V.L.P. *Designing and Conducting Mixed Methods Research*; Sage Publication: London, UK, 2007.

27. Hesse-Biber, S.N. *Mixed Methods Research: Merging Theory with Practice*; Guilford Press: London, UK, 2010.

28. Thompson, P. *Voice of the Past: Oral History*; Oxford University Press: Oxford, UK. 2000.

29. Conway, D. Managing water in the Ruaha River Basin: The role of climate variability and risk. In Proceedings of the 2005 East African Integrated River Basin Management Conference, Morogoro, Tanzania, 7–9 March 2005; pp. 420–427.

30. Keskinen M.; Chinvanno, S.; Kummu, M.; Nuorteva, P.; Snidvongs, A.; Varis, O.; Västilä, K. Climate change and water resources in the Lower Mekong River Basin: Putting adaptation into the context. *J. Water Clim. Change* **2010**, *1*, 103–117.

31. Lovell, C.; Mandondo, A.; Moriarty, P. The question of scale in integrated natural resource management. In *Integrated Natural Resource Management. Linking Productivity, the Environment and Development*; Campbell, B.M., Sayer, J., Eds.; CABI: Wallingford, UK, 2003; pp. 109–136.

32. Domroes, M.; Schaefer D. Trends of recent temperature and rainfall changes in Sri Lanka. In Proceedings of 2000 International Conference on Climate Change and Variability, Tokyo, Japan, 13–17 September 2000; pp. 197–204.

33. Zubair, L.; Siriwardhana, M.; Chandimala, J.; Yahiya, Z. Predictability of Sri Lankan rainfall based on ENSO. *Int. J. Climatol.* **2008**, *28*, 91–101.

34. Manawadu, L.; Fernando, N. *Climate Changes in Sri Lanka*; Department of Meteorology, Sri Lanka: Colombo, Sri Lanka, 2008.

35. Muththuwatta, L.P.; Liyanage, P.K.N.C. Climate Change Models Shift Boundaries of Agro-Ecological Zones in Sri Lanka. Available online: http://climatenet.blogspot.de/2013/09/climate-change-models-shift-boundaries.html (accessed on 10 July 2014).

36. Shantha, W.W.A.; Jayasundara, J.M.S.B. Study on changes of rainfall in the Mahaweli Upper Watershed in Sri Lanka due to climatic changes and develop a correction model for global warming. In Proceedings of the 2005 International Symposium on the Stabilisation of Greenhouse Gas Concentrations, Exeter, UK, 1–3 February 2005.

37. Eriyagama, N.; Smakhtin, V.; Chandrapala, L.; Fernando, K. *Impacts of Climate Change on Water Resources and Agriculture in Sri Lanka: A Review and Preliminary Vulnerability Mapping*; IWMI Research Report 135; International Water Management Institute: Colombo, Sri Lanka, 2010.

38. Cruz, R.V.; Harasawa, H.; Lal, M.; Wu, S.; Anokhin, Y.; Punsalmaa, B.; Honda, Y.; Jafari, M.; Li, C.; Huu Ninh, N. Asia. In *Climate Change 2007: Impacts, Adaptation and Vulnerability*; Parry, M.L., Canziani, O.F., Palutikof, J.P., Van der Linden, P.J., Hanson, C.E., Eds.; Cambridge University Press: Cambridge, UK, 2007; pp. 469–506.

39. Ministry of Finance and Planning, Sri Lanka. Economic Perspective: Sri Lanka. In *Annual Report 2012*; Ministry of Finance and Planning, Sri Lanka: Colombo, Sri Lanka, 2012; pp. 19–48.

40. IPCC. Summary for policymakers. In *Climate Change 2014: Impacts, Adaptation, and Vulnerability. Part A: Global and Sectoral Aspects. Contribution of Working Group II to the Fifth Assessment Report of the Intergovernmental Panel on Climate Change*; Field, C.B., Barros, V.R., Dokken, D.J., Mach, K.J., Mastrandrea, M.D., Bilir, T.E., Chatterjee, M., Ebi, K.L., Estrada, Y.O., Genova, R.C., *et al.*, Eds.; Cambridge University Press: Cambridge, UK, 2014; pp. 1–32.

41. Jayawardene H.K.W.I.; Sonnadara, D.U.J.; Jayewardene, D.R. Trends of rainfall in Sri Lanka over the last century. *Sri Lankan J. Phys.* **2005**, *6*, 7–17.

42. Zubair L. ENSO influences on Mahaweli Stream flow in Sri Lanka. *Int. J. Climatol.* **2003**, *23*, 91–102.

43. Dharmaratne, G.H.P. Impact of changing weather patterns. *Vidurava* **2005**, *22*, 9–13.

44. Riedlinger, D.; Berkes, F. Contributions of traditional knowledge to understanding climate change in the Canadian Arctic. *Polar Record* **2001**, *37*, 315–328.

45. Nyong, A.; Adesina, F.; Elasha, B.O. The value of indigenous knowledge in climate change mitigation and adaptation strategies in the African Sahel. *Mitig. Adapt. Strat. Glob. Change* **2007**, *12*, 787–797.

46. David, W.; Ploeger, A. Indigenous Knowledge (IK) of water resources management in West Sumatra, Indonesia. *Fut. Food: J. Food Agric. Soc.* **2014**, *2*, 52–60.

47. Byg, A.; Salick, J. Local perspectives on a global phenomenon—Climate change in Eastern Tibetan villages. *Glob. Environ. Change* **2009**, *19*, 156–166.

48. Mahaweli Authority. *Performance-2012 and Investment Plan–2013, Performance Report up to August-2012*; Planning and Monitoring Unit in Mahaweli Authority: Colombo, Sri Lanka, 2012.

49. The Disaster Management Centre (DMC) of the Ministry of Disaster Management. *Sri Lanka National Report on Disaster Risk, Poverty and Human Development Relationship*; The Disaster Management Centre (DMC) of the Ministry of Disaster Management: Colombo, Sri Lanka, 2014.

50. Mahaweli Authority. *Mahaweli Hand Book (Statistical Hand Book) 2011–2012*; Planning and Monitoring Unit in Mahaweli Authority: Colombo, Sri Lanka, 2012.

51. Parliament of Sri Lanka. *Mahaweli Authority Parliament—Act No. 23 of 1979*; Parliament of Sri Lanka: Sri Jayawardenapura, Sri Lanka, 1979.

52. Hettige, S.T. Socioeconomic issues relating to sustainability of the mahaweli upper catchment Sri Lankan. *J. Soc. Sci.* **1997**, *20*, 43–62.

53. Lakshman, W.D.; Tisdell, C.A. Introduction to Sri Lanka's development since independence. In *Sri Lanka's Development Since Independence: Socio-Economic Perspectives and Analyses*; Nova Science Publishers Inc.: New York, NY, USA, 2000; pp. 1–22.

54. Haturusinha, R.L.; Theivasagayam, K.; Perera, D.T.D. Water and sanitation for Mahaweli. In Proceedings of the 1994 WEDC Conference, Colombo, Sri Lanka, 22–26 August 1994; pp. 242–245.

55. Wettasinha, C. Scaling up participatory development in agricultural settlements. *Leisa. Mag.* **2001**, 39–42.

56. Perera, G.D.; Sennema, B. Towards sustainable development in Mahaweli Settlements through farmer participation. In *Participatory Technology Development Working Paper 6*; Diop, J.-M., Waters-Bayer, A., Eds.; Ecoculture Netherlands and Mahaweli Authority: Colombo, Sri Lanka, 2002; pp. 1–23.

57. Hewawasam, T. Effect of land use in the upper Mahaweli catchment area on erosion, landslides and siltation in hydropower reservoirs of Sri Lanka. *J. Nat. Sci. Found. Sri Lanka* **2010**, *38*, 3–14.

58. Lezama Escalante, C. A Cultural approach to water management policies. In *Five Years of EXCEED. Sustainable Water Management in Developing Countries*; Bahadir, M., Haarstrick, A., Eds.; Technische Universität Braunschweig: Braunschweig, Germany 2014; pp. 256–265.

59. AQUASTAT. Available online: http://www.fao.org/nr/water/aquastat/main/index.stm. (accessed on 20 October 2014).

60. WWAP (United Nations World Water Assessment Programme). *The United Nations World Water Development Report: Water and Energy*; UNESCO: Paris, France, 2014.

61. Bastiaassen, W.G.M.; Chandrapala, L. Water balance availability across Sri Lanka for assessing agricultural and environmental water use. *Agric. Water Manag.* **2003**, *58*, 171–192.

62. Central Bank of Sri Lanka. *Annual Report 2012*; Central Bank of Sri Lanka: Colombo, Sri Lanka, 2012.

63. Central Bank of Sri Lanka. *Annual Report 2011*; Central Bank of Sri Lanka: Colombo, Sri Lanka, 2011.

64. Wickremasekara, D. Paddy Cultivation Slashed; Rice to be Imported. Available online: http://www.sundaytimes.lk/140406/news/paddy-cultivation-slashed-rice-to-be-imported-91848.html (accessed on 20 July 2014).

65. Malmgren, B.A.; Hulugalla, R.; Hayashi, Y.; Mikami, T. Precipitation trends in Sri Lanka since the 1870s and relationships to El Niño—Southern oscillation. *Int. J. Climatol.* **2003**, *23*, 1235–1252.

66. Niranjan, F.; Jayatilaka, W.; Singh, N.P.; Bantilan, M.C.S. *Mainstreaming Grassroots Adaptation and Building Climate Resilient Agriculture in Sri Lanka*; Policy Brief No. 20; International Crops Research Institute for the Semi-Arid Tropics: Hyderabad, India, 2013.

67. Matthews, R.B.; Kropff, M.J.; Horie, T.; Bachelet, D. Simulating the impact of climate change on rice production in Asia and evaluating options for adaptation. *Agric. Syst.* **1997**, *54*, 399–425.

68. Welch, J.R.; Vincent, J.R.; Auffhammer, M.; Moya, P.F.; Dobermann, A.; Dawe, D. Rice yields in tropical/subtropical Asia exhibit large but opposing sensitivities to minimum and maximum temperatures. *Proc. Natl. Acad. Sci. USA* **2010**, *107*, 14562–14567.

69. Sri Lanka Disaster Knowledge Network. Disaster Profile of Sri Lanka. Available online: http://www.saarc-sadkn.org/countries/srilanka/disaster_profile.aspx (accessed on 10 October 2014).

70. Geekiyanage, N.; Pushpakumara, D.K.N.G. Ecology of ancient Tank Cascade Systems in island Sri Lanka. *J. Mar. Island Cult.* **2013**, *2*, 93–101.

71. Madduma Bandara, C.M. Tank cascade systems in Sri Lanka: Some thoughts on their development implications. In *Summaries of Papers Presented at Irrigation Research Management Unit Seminar Series during 1994*; Haq, K.A., Wijayaratne, C.M., Samarasekera B.M.S., Eds.; IIMI (International Irrigation Management Institute): Colombo, Sri Lanka, 1995; p. 14.

72. Schütt, B.; Bebermeier, W.; Meister, J.; Withanachchi, C.R. Characterisation of the Rota Wewa tank cascade system in the vicinity of Anuradhapura, Sri Lanka. *J. Geogr. Soc. Berl.* **2013**, *144*, 51–68.

73. Withanachchi, C.R. Hydro-technology in dry zone irrigation management of ancient Sri Lanka and its potential benefits for the world the international conference on Traditional Knowledge for Water Resource Management (TKWRM). In Proceedings of the 2012 International Center on Qanats and Historic Hydraulic Structure (ICQHS), Yazd, Iran, 21–23 February 2012; pp. 1–7.

74. Panapitiya, M. "Inconvenient Truth" behind engineering designs of irrigation projects developed during the last century. *Econ. Rev. Peopl. Bank Colombo* **2010**, *36*, 16–20.

75. Gunawardena, E.R.N. Operationalizing IWRM Through River Basin Planning and Management. Available online: http://tinyurl.com/ky53p9p (accessed on 5 October 2013).

76. Merrey, D.J. Potential for devolution of management to farmers' organizations in an hierarchical irrigation management agency: The case of the Mahaweli Authority of Sri Lanka. In *The Blurring of a Vision—The Mahaweli: Its Social, Economic and Political Implications*; Müller, H.P., Hettige, S.T., Eds.; Sarvodaya Publication: Ratmalana, Sri Lanka, 1995; pp. 222–243.

77. Müller, H.P.; Hettige, S.T. Introduction the Mahaweli-The Blurring of a Vision. In *The Blurring of a Vision—The Mahaweli: Its Social, Economic and Political Implications*; Müller, H.P., Hettige, S.T., Eds.; Sarvodaya Publication: Ratmalana, Sri Lanka, 1995; pp. 1–22.

78. Mollinga, P.P. The water resources policy process in India: Centralization, polarization and new demands on governance. In *Governance of Water: Institutional alternatives and Political Economy*; Ballabh, V., Ed.; Sage: New Delhi, India, 2008; pp. 339–370.

79. Wijesekera, N.T.S.; Wickramaarachchi, T.N. Reality of irrigation water use and suggestions for better management: A comparison of two schemes from Sri Lanka. *Water Sci. Technol.* **2003**, *48*, 197–206.

Assessing the Impacts of Sea Level Rise on Salinity Intrusion and Transport Time Scales in a Tidal Estuary, Taiwan

Wen-Cheng Liu [1,2],* and Hong-Ming Liu [1]

[1] Department of Civil and Disaster Prevention Engineering, National United University, Miaoli 36003, Taiwan; E-Mail: dslhmd@gmail.com

[2] Taiwan Typhoon and Flood Research Institute, National Applied Research Laboratories, Taipei 10093, Taiwan

* Author to whom correspondence should be addressed; E-Mail: wcliu@nuu.edu.tw; wcliu@narlabs.org.tw.

Abstract: Global climate change has resulted in a gradual sea level rise. Sea level rise can cause saline water to migrate upstream in estuaries and rivers, thereby threatening freshwater habitat and drinking-water supplies. In the present study, a three-dimensional hydrodynamic model was established to simulate salinity distributions and transport time scales in the Wu River estuary of central Taiwan. The model was calibrated and verified using tidal amplitudes and phases, time-series water surface elevation and salinity distributions in 2011. The results show that the model simulation and measured data are in good agreement. The validated model was then applied to calculate the salinity distribution, flushing time and residence time in response to a sea level rise of 38.27 cm. We found that the flushing time for high flow under the present condition was lower compared to the sea level rise scenario and that the flushing time for low flow under the present condition was higher compared to the sea level rise scenario. The residence time for the present condition and the sea level rise scenario was between 10.51 and 34.23 h and between 17.11 and 38.92 h, respectively. The simulated results reveal that the residence time of the Wu River estuary will increase when the sea level rises. The distance of salinity intrusion in the Wu River estuary will increase and move further upstream when the sea level rises, resulting in the limited availability of water of suitable quality for municipal and industrial uses.

Keywords: sea level rise; climate change; salinity intrusion; flushing time; residence time; model simulation; hydrodynamics; Wu River estuary

---

1. Introduction

Global warming is irrefutably causing sea level to rise. The global mean sea level raised by ~20 cm, along with a rise in the regional mean sea level, as the global air temperature increased by ~0.5–0.6 °C during the 20th century [1,2]. In Taiwan, the surface temperature has raised approximately 1.0–1.4 °C over the last 100 years [3]. Over the past 80 years, the annual precipitation has increased in northern Taiwan and declined in central and southern Taiwan [4]. The changing climate has also caused some impacts on river ecosystems in Taiwan; more-frequent habitat disturbances have caused both a shift in aquatic organism distributions and population decline [5].

Sea level rise can cause saline water to migrate upstream to points where freshwater previously existed [6]. Several studies indicated that sea level rise would increase the salinity in estuaries [7,8], which would result in changes in stratification and estuarine circulation [9]. Salinity migration could cause shifts in salt-sensitive habitats and could thus affect the distribution of flora and fauna.

Salinity intrusion may decrease the water quality in an estuary, so that its water becomes unsuitable for certain uses, such as agricultural, industrial and drinking purposes. Therefore, the determination of the salinity distribution along an estuary is a major interest for water managers in estuaries and coastal regions. The evaluation of transport time scales is highly related to the water quality and ecological health of different aquatic systems [10].

Several numerical modeling studies have shown that increases in sea level have impacts on estuarine salinity. Hull and Tortoriello [11] used a one-dimensional model to estimate the impacts of sea level rise and found that a sea level rise of 0.13 m would result in a salinity increase of 0.4 psu (practical salinity unit) in the upper portion of the Delaware Bay during low-flow periods. Grabemann et al. [12] simulated a 2-km upstream advance of the brackish water zone for a sea level rise of 0.55 m in the Weser Estuary, Germany. Hilton et al. [7] found an average salinity increase of approximately 0.5 with a 0.2 m sea level rise based on model simulations in Chesapeake Bay. Chua et al. [13] found that the intrusion of salt water into San Francisco Bay and the flushing rate both increase as the sea level rises. Bhuiyan and Dutta [8] applied a one-dimensional model to investigate the impact of sea level rise on river salinity in the Gorai River network and found that a sea level rise of 0.59 m increased salinity by 0.9 at a distance of 80 km upstream of the river mouth. Rice et al. [14] concluded that salinity in the James River would intrude about 10 km farther upstream for a sea level rise of 1.0 m using a three-dimensional hydrodynamic model.

Numerous studies have reported the influences of sea level rise on estuarine salinity, stratification, exchange flow, residence time, material transport processes and other relevant processes in estuaries [8,9,14]. However, the reports regarding the impacts of sea level rise on salinity intrusion and transport time scales have not yet been studied in Taiwan's estuaries. The objective of the present study is to examine the salinity intrusion, flushing time and residence time in response to sea level rise in the Wu River estuary of central Taiwan using a three-dimensional hydrodynamic and salinity transport model. The model was validated with observed amplitudes and phases, water levels and salinity to ascertain the model's accuracy and capability. The model was then applied to the Wu River estuary to calculate the salinity distributions and transport time scales based on sea level rise projections. The model results were used to investigate how sea level rise affects salinity intrusion, flushing time and residence time in Taiwan's Wu River estuary.

2. Study Area

The Wu River system is the most important river in central Taiwan (Figure 1a). The mean tidal range at the mouth of the Wu River is 3.8 m above mean sea level. Tidal propagation is the dominant mechanism controlling the water surface elevation. The M_2 (principal lunar semi-diurnal) tide is the primary tidal constituent at the mouth of the Wu River [15]. The main tributaries are the Fazi River, Dali River, Han River and Maoluo River. The downstream reaches of the main Wu River are affected by tides, whereas the tributaries are not subject to tidal effects and are therefore not affected by salt water intrusion. The drainage basin of the Wu River, which is the fourth-largest river basin in Taiwan, covers approximately 2026 km^2. The total channel length is 117 km, and the mean channel slope is 1/92. The morphology of the Wu River displays different features in each segment, molded by natural forces, as well as anthropogenic activities exerted upon the paleo-riverbed built ages ago. The riverbed is composed of silt and sand in the estuary. The mean annual precipitation in this region is 2087 mm. The ample flow season is from May to September, accounting for 70% of the river discharge, and the dry season is from January to February. The daily flow data from 1969 to 2011 at the Dadu Bridge, collected by the Water Resources Bureau of Taiwan, are analyzed in this study. The data analysis indicates that the Q_{75} low flow is 41.2 m^3/s. The definition of Q_{75} is the flow that is equaled or exceeded for 75% of the time. The river, which flows into the Taiwan Strait, is located in a temperate area characterized by intense agricultural and industrial activities. The Wu River catchment is also an important water supply source for central Taiwan. Figure 1b shows the topography of the Wu River estuary and its adjacent coastal sea. This figure indicates that the greatest depth in the study area is 70 m (below mean sea level) near the corner of the coastal sea.

Figure 1. (a) Map of the Wu River system and **(b)** bathymetry of the Wu River estuary and adjacent coastal sea.

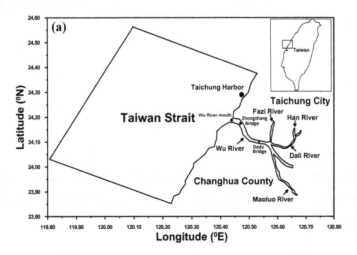

Figure 1. *Cont.*

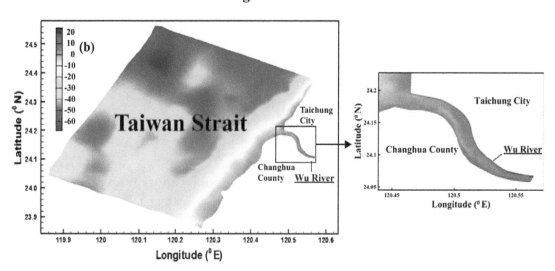

3. Materials and Methods

3.1. Sea Level Rise Projection

The existence of sea level rise is undeniable. Church and White [16] estimated the global mean sea level rise rates from tidal gauges and satellite altimetry as follows: 1.7 ± 0.2 mm year^{-1} for 1900–2009 and 1.9 ± 0.4 mm year^{-1} for 1961–2009, both of which are comparable to the ~1.8 mm year^{-1} rate obtained from GPS-derived crustal velocities and tidal gauges around North America [17]. Global tide gauge records, satellite data and modeling provide mean historical rates of ~1.7–1.8 mm year^{-1} for the 20th century and ~3.3 mm year^{-1} for the last few decades [18].

The purpose of this study is to identify the response of the Wu River estuary to potential future sea level rise based on the analyzed results of observed sea level. Tseng *et al.* [19] investigated the pattern and trends of sea level rise in the region seas around Taiwan through the analyses of long-term tide-gauge and satellite altimetry data. They found that consistent with the coastal tide-gauge records, satellite altimetry data showed similar increasing rates (+5.3 mm/year) around Taiwan. They did not include wave breaking around the river mouth and coastal seas, resulting in water level rise due to wave set-up. In this study, the wave set-up issue also did not take into account. The linear regression method was used to yield the sea level rise trend according to the monthly average water surface elevation collected from 1971 to 2011 at the Taichung Harbor station, which is shown in Figure 2. The equation of linear regression can be expressed as:

$$Y = 0.43X - 591.8 \tag{1}$$

where X is the time (year) and Y is the sea water level (cm). We found that the rate of sea level rise was 4.3 mm/year at the Taichung Harbor station. Huang *et al.* [20] estimated the sea level rise at the Taichung Harbor station using the data of tidal gauge and satellite altimetry and found that the rate of sea level rise was 3.7 mm/year. Their results are similar to our estimation on the rate of sea level rise. The sea level rise in 2011 was set up zero to project the sea level rise in 2010. The future projected sea level rise of 38.27 cm in 2100 was used in the simulation scenario.

Figure 2. Linear regression for the sea level rise trend at the Taichung Harbor station.

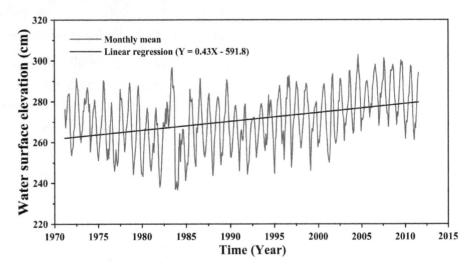

3.2. Three-Dimensional Hydrodynamic Model

A three-dimensional, semi-implicit Euler-Lagrange finite-element model (SELFE) [21] was implemented to simulate the hydrodynamics and salinity transport in the Wu River estuary and its adjacent coastal sea. SELFE solves the Reynolds-stress averaged Navier-Stokes equations, which consist of the conservation laws for mass and momentum and the use of the hydrostatic and Boussinesq approximations, yielding the following free-surface elevation and three-dimensional water velocity equations:

$$\frac{\partial u}{\partial x}+\frac{\partial v}{\partial y}+\frac{\partial w}{\partial z}=0 \tag{2}$$

$$\frac{\partial \eta}{\partial t}+\frac{\partial}{\partial x}\int_{H_R-h}^{H_R+\eta}udz+\frac{\partial}{\partial y}\int_{H_R-h}^{H_R+\eta}vdz=0 \tag{3}$$

$$\frac{\partial u}{\partial t}+u\frac{\partial u}{\partial x}+v\frac{\partial u}{\partial y}+w\frac{\partial u}{\partial z}=fv-\frac{\partial}{\partial x}\left\{g(\eta-\alpha\varphi)+\frac{P_a}{\rho_o}\right\}$$
$$-\frac{g}{\rho_o}\int_z^{H_R+\eta}\frac{\partial \rho}{\partial x}dz+\frac{\partial}{\partial z}(v\frac{\partial u}{\partial z}) \tag{4}$$

$$\frac{\partial v}{\partial t}+u\frac{\partial v}{\partial x}+v\frac{\partial v}{\partial y}+w\frac{\partial v}{\partial z}=-fu-\frac{\partial}{\partial y}\left\{g(\eta-\alpha\varphi)+\frac{P_a}{\rho_o}\right\}$$
$$-\frac{g}{\rho_o}\int_z^{H_R+\eta}\frac{\partial \rho}{\partial y}dz+\frac{\partial}{\partial z}(v\frac{\partial v}{\partial z}) \tag{5}$$

$$\frac{DS}{Dt}=\frac{\partial}{\partial z}(K_v\frac{\partial S}{\partial z})+F_s \tag{6}$$

$$\rho=\rho_0(p,S) \tag{7}$$

where (x,y) are the horizontal Cartesian coordinates; (ϕ,λ) are the latitude and longitude, respectively; z is the vertical coordinate, positive upward; t is time; H_R is the z-coordinate at the reference level (mean sea level); $\eta(x,y,t)$ is the free-surface elevation; $h(x,y)$ is the bathymetric depth; u,v and w are the velocities in the x, y and z directions, respectively; f is the Coriolis force; g is the acceleration of gravity; $\varphi(\phi,\lambda)$ is the tidal potential; α is the effective earth elasticity factor (=0.69); $\rho(\vec{x},t)$ is the water density, of which the default reference value; ρ_o, is set to 1,025 kg/m^3; $P_a(x,y,t)$ is the atmospheric pressure at the free surface; p is the pressure; v is the vertical eddy viscosity; s is the salinity; K_v is the vertical eddy diffusivity for salinity and F_s is the horizontal diffusion for the transport equation.

The vertical boundary conditions for the momentum equation, especially the bottom boundary condition, play an important role in the SELFE numerical formulation, as it involves the unknown velocity. In fact, as a crucial step in solving the differential system, SELFE uses the bottom condition to decouple free-surface Equation (3) from momentum Equations (4) and (5). The vertical boundary conditions for the momentum equation are presented as below.

At the water surface, the balance between the internal Reynolds stress and the applied shear stress yields:

$$v\frac{\partial \vec{u}}{\partial z} = \vec{\tau_w} \ \text{ at } \ z = \eta \tag{8}$$

where the specific stress, τ_w, can be parameterized using the approach [22].

The boundary condition at the bottom plays an important role in the SELFE formulation, as it involves unknown velocity. Specifically, at the bottom, the no-slip condition $(U=V=0)$ is usually replaced by a balance between the internal Reynolds stress and the bottom frictional stress, i.e.,

$$v\frac{\partial \vec{u}}{\partial z} = \tau_b \ \text{ at } \ z = -h \tag{9}$$

where the bottom stress is $\vec{\tau_b} = C_D |u_b| \vec{u_b}$.

The velocity profile inside the bottom boundary layers obeys the logarithmic law:

$$\vec{u} = \frac{\cdots \mathsf{L}(\cdots \cdots) \cdots \mathsf{u} \mathsf{J}}{\ln(\delta_b / z_0)} \vec{u_b}, \ (z_0 - h \le z \le \delta_b - h) \tag{10}$$

which is subject to be smoothly matched to the exterior flow. In Equation (10), δ_b is the thickness of the bottom computational cell; z_0 is the bottom roughness, which is determined through model calibration and verification; and u_b is the bottom velocity, measured at the top of the bottom computational cell. The Reynolds stress inside the boundary layer is derived from Equation (11) as:

$$v\frac{\partial \vec{u}}{\partial z} = \frac{v}{(z+h)\ln(\delta_b / z_0)} \vec{u_b} \tag{11}$$

The SELFE model uses the Generic Length Scale (GLS) turbulence closure approach of Umlauf and Burchard [23], which has the advantage of incorporating most of the 2.5-equation closure model. The SELFE model treats advection in the momentum equation using a Euler–Lagrange methodology. A detailed description of the turbulence closure model, the vertical boundary conditions for the momentum equation and the numerical solution methods can be found in Zhang and Baptista [21].

3.3. Computation of Flushing Time

The flushing time can be conveniently determined by the freshwater fraction approach [24–27], which can be determined from salinity distributions. This technique provides an estimation of the time scale over which contaminants and/or other material released in the estuary are removed from the system. Using the freshwater fraction method, the flushing time in an estuary can be expressed as:

$$T_f = \frac{F}{Q} = \frac{\int\limits_{vol} f \cdot d(V)}{Q} \tag{12}$$

where F is the accumulated freshwater volume in the estuary, which can be calculated by integrating the freshwater volume; $d(V)$, in all the sub-divided model grids over a period of time. In estuaries with unsteady river flow and tidal variations; F and Q are the approximate average freshwater volume and average freshwater input, respectively, over several tidal cycles for a period of time, such as a week or a month [20,21]. The term, f, is the freshwater content or the freshwater fraction, which is described by:

$$f = \frac{S_0 - S}{S_0} \tag{13}$$

where S_0 is the salinity in the ocean; and S is the salinity at the study location.

3.4. Computation of Residence Time

The time scales associated with the residence time of water parcels and their associated dissolved and suspended materials in a specific water body due to different transport mechanisms (*i.e.*, advection and dispersion) are fundamental physical characteristics of that water body. Residence time is defined as the time required for a water parcel to leave the region of interest for the first time [28]. Several methodologies for the computation of residence time have been reported in the literature [29–34]. In the present study, the computational method follows the procedures outlined by Takeoka [30]. Consider that a region of interest contains a finite mass of tracer given by $M(0)$ at the initial time $t = t_0$. If we define the remaining mass of tracer at a certain time, t, within the system as $M(t)$, the distribution function of the residence time can be defined as:

$$T_r = -\frac{1}{M(0)} \frac{dM(t)}{dt} \tag{14}$$

where T_r is the distribution function of residence time. The total mass of the tracer will completely leave the system at a given moment when $\lim_{t \to \infty} M(t)$ is equal to zero. The average residence time of the tracer can be computed by:

$$\overline{T_r} = \int_{t_0}^{\infty} t T_r(t) dt = \int_{t_0}^{\infty} \frac{M(t)}{M(0)} dt \tag{15}$$

The fraction of mass $r(t) = M(t) / M(0)$ is known as the remnant function. Note that $M(t)$, the mass of the tracer that remains in the region of interest at a certain time; t, can be computed numerically based on the tracer concentration by:

$$M(t) = \int C(x,t) dV \tag{16}$$

where $C(x,t)$ is the tracer concentration in a differential volume ; dV, at a given time; t, and position, x, within the system. It is expected that a mass of tracers injected close to the boundaries of a given region has a lower residence time than does the residence time of tracers injected at the center of such a region.

3.5. Model Schematization

An accurate representation of the bottom topography in the model grid is critical for successful estuarine, coastal and ocean modeling. In this study, the bottom topography data in the coastal seas and Wu River estuary were obtained from the databank of the National Science Council and the Water Resources Agency in Taiwan, respectively. The modeling domain in the horizontal plane covers an area of 60 km × 45 km at the coastal sea boundary. Because SELFE uses a combination of Eulerian–Lagrangian and implicit time stepping, it does not have to satisfy the usual Courant–Friedrich–Levy (CFL) constraint for numerical stability [21]. However, 120 s was chosen as the time step (Δt). Trial-and-error tests with other time steps demonstrated that the model results did not improve significantly with lower values. The model meshes for the Wu River estuary and the coastal sea consisted of 3541 polygons and 1974 grids, respectively (Figure 3). Because the model domain covers deep bathymetry in the coastal sea and shallow bathymetry near the coastline, ten levels, varying in thickness from 0.2 to 7 m, were adopted for vertical discretization in the SELFE model.

Figure 3. Unstructured grids in the modeling domain.

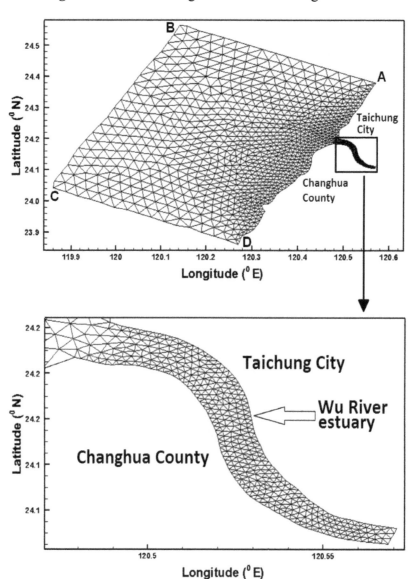

4. Model Calibration and Verification

To ascertain the model accuracy for applications on the assessment of sea level rise on salinity intrusion and transport time scales, a set of observational data collected in 2011 were used to calibrate and verify the model and to validate its capability to predict amplitudes and phases, water surface elevation and salinity distribution.

4.1. Calibration with Amplitudes and Phases

The local bottom roughness height (z_o) is similar to the Manning coefficient, affecting the water level calculations for the coastal sea and estuary. The values of local bottom roughness height were iteratively adjusted by trial and error until the simulated and observed tidal levels were satisfactory [35]. In this study, the bottom roughness was adjusted to calibrate the amplitudes and phases at Taichung Harbor. The model calibration of the amplitudes and phases was conducted using measured data on the daily freshwater discharge at the Dadu Bridge in 2011. A five-constituent tide (*i.e.*, M_2, S_2, N_2, K_1 and O_1) was adopted in the model simulation as a forcing function at the coastal sea boundaries (shown in Table 1). Because the amplitudes of fourth-diurnal, such as M_4 (first overtide of M_2 constituent) and MS_4 (a compound tide of M_2 and S_2), comparing to diurnal and semi-diurnal tides, were relatively small, a five-constituent tide was used to force the open boundaries only. The amplitudes and phases of these five tidal constituents were used to generate time-series water surface elevations along the open boundaries. The freshwater discharge inputs from Dadu Bridge in 2011 are shown in Figure 4. The maximum freshwater discharge reached 690 m^3/s during the typhoon event.

The model simulation was run for one year in 2011. Harmonic analysis was performed on the time series of the model simulated water surface elevation at Taichung Harbor. The bottom roughness height was adjusted carefully, and the results are presented in Figure 5. The results show the comparison of the amplitudes and phases of harmonic constants between computed and observed tides. The differences between the computed and observed tidal constituents for amplitude and phase are in the range of 0.01–0.02 m and 0.45°–4.21°, respectively. The differences in amplitude and phase are quite small.

Table 1. The amplitudes and phases used for the model simulation at the coastal sea boundaries.

Constituent	Boundary at Point A		Boundary at Point B		Boundary at Point C		Boundary at Point D	
	Amplitude (m)	Phase (°)	Amplitude (m)	Phase (°)	Amplitude (m)	Phase (°)	Amplitude (m)	Phase (°)
M_2	1.82	266.06	1.88	266.91	1.60	272.39	1.57	267.23
S_2	0.51	14.45	0.53	16.34	0.44	26.29	0.43	20.55
N_2	0.25	28.89	0.26	29.77	0.22	36.02	0.21	31.15
K_1	0.27	161.62	0.29	160.20	0.29	167.21	0.28	167.16
O_1	0.21	279.53	0.22	277.24	0.23	283.55	0.22	284.03

Notes: boundaries at Points A, B, C and D are shown in Figure 3; M_2 is principal lunar semi-diurnal constituent; S_2 is principal solar semi-diurnal constituent; N_2 is larger lunar elliptic semi-diurnal constituent; K_1 is luni-solar declinational diurnal constituent; and O_1 is lunar declinational diurnal constituent.

Figure 4. Freshwater discharge inputs at the Dadu Bridge in 2011.

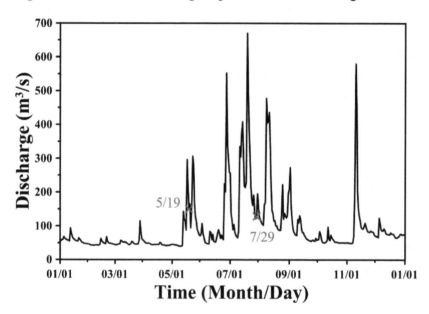

Figure 5. Comparisons of amplitude and phase of five major tidal harmonics computed with a three-dimensional model and obtained from tide measurements (**a**) amplitude; (**b**) phase.

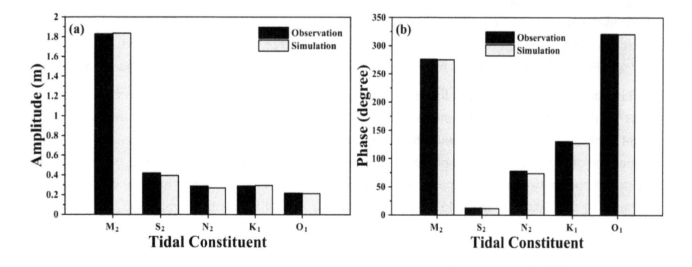

4.2. Verification of Water Surface Elevation

After calibrating the amplitudes and phases, the time-series data of observed water surface elevation were used to verify the model. Figure 6 presents the verified results for water surface elevations at Taichung Harbor station in May and July, 2011. The mean absolute errors of the differences between the measured hourly water levels and the computed water levels for 11–21 May and 22–31 July were 0.147 m and 0.157 m, respectively. The corresponding root-mean-square errors were 0.183 m and 0.193 m, respectively. These results demonstrate that the model can accurately predict the water surface elevation for varying river discharge input and tidal forcing at coastal sea boundaries. A constant bottom roughness height (z_0 = 0.01 cm) was adopted in the model for calibration and verification.

Figure 6. Comparison of model results and observed water surface elevation during the periods of (**a**) 11–21 May 2011 and (**b**) 22–31 July 2011 at the Taichung Harbor station.

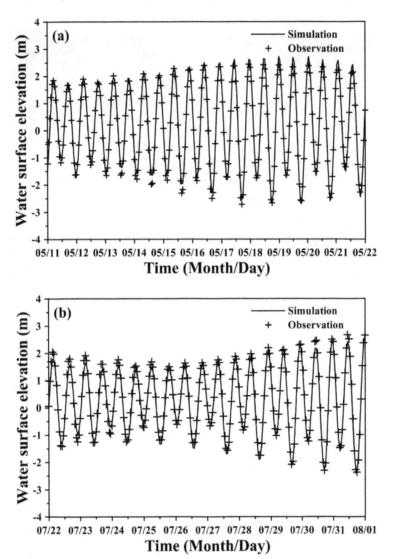

4.3. Calibration and Verification of Salinity Distribution

Salinity distributions reflect the combined results of all processes, including density circulation and mixing processes [36]. In the present study, the salinity distributions were measured *in situ* using conductivity-temperature-depth equipment at six locations in the Wu River estuary during the flood tide surveys. The salinities at four vertical layers of each station in the water column were measured and were then used for model calibration and verification. The salinities of open boundaries in the coastal sea were set to a constant value (*i.e.*, 35 psu). The upstream boundary at the Dadu Bridge was also specified with daily freshwater discharges, and the salinity was set to 0 psu. Figures 7 and 8 present the comparisons of measured and simulated salinity distributions on 19 May and 29 July, 2011, for model calibration and verification purposes, respectively. The freshwater discharges on 19 May and 29 July 2011, were 149.3 m³/s and 127.47 m³/s, respectively (shown in Figure 4). The results show that the model-computed salinity distributions agree well with the field observations. The root-mean-square errors for 19 May and 29 July, 2011, were 0.75 psu and 0.53 psu, respectively.

Figure 7. Comparison of salinity distribution along the Wu River estuary. (**a**) measurements and (**b**) model simulation on 19 May 2011, for model calibration.

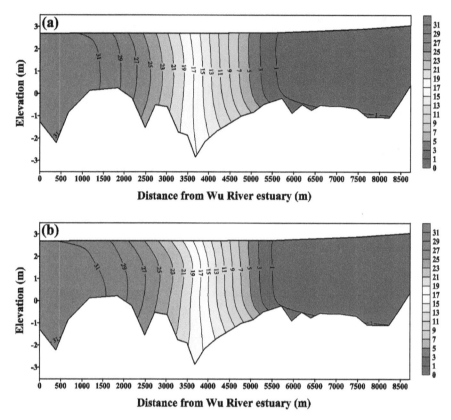

Figure 8. Comparison of salinity distribution along the Wu River estuary. (**a**) measurements and (**b**) model simulation on 29 July 2011 for model verification.

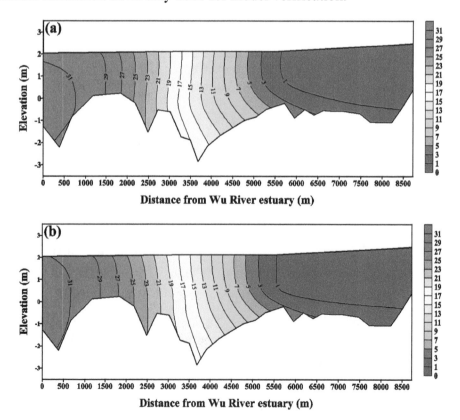

5. Results and Discussion

The validated three-dimensional hydrodynamic model was used to calculate the salinity distribution and transport time scale response to different discharges with sea level rise scenarios and without sea level rise (*i.e.*, the present condition) in the Wu River estuary. Figure 9 presents flow duration curves at the Dadu Bridge. The daily flow data from 1969 to 2011, collected by the Central Water Resources Bureau of Taiwan, were analyzed. The freshwater discharges with Q_{10} (the flow that is equaled or exceeded 10% of time) to Q_{90} flow conditions are listed in Table 2. For the cases of Q_{10} and Q_{90} flows, the discharges at the Dadu Bridge are 229.0 and 26.0 m^3/s, respectively. Five tidal constituents (M_2, S_2, N_2, K_1 and O_1) were specified to generate a time-series of water surface elevation as the open boundary conditions at the coastal sea for model simulation. A constant salinity of 35 psu at the open boundaries was used for model simulation. A future sea level rise of 38.27 cm in 2100 was used for the model simulation scenario.

Figure 9. Flow duration curve at Dadu Bridge.

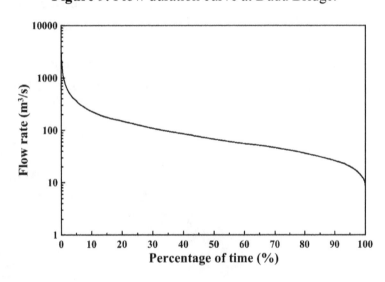

The influence of wind on estuarine circulation has been recognized for many years [37]. In a shallow estuary, the residence time can vary in response to variations in wind-induced flushing [38]. However, in the present study, the wind forcing was excluded in the model simulation for calculating transport time scales with sea level rise scenarios and without sea level rise (*i.e.*, the present condition).

Table 2. Freshwater discharge at upstream boundaries for the model simulation.

Freshwater discharge	Flow rate at Dadu Bridge (m^3/s)
Q_{10}	229.0
Q_{20}	149.0
Q_{30}	108.7
Q_{40}	85.6
Q_{50}	67.5
Q_{60}	55.5
Q_{70}	46.5
Q_{80}	36.5
Q_{90}	26.0

5.1. Sea Level Rise Effects on Salinity Distribution

To quantify the spatial and vertical variations in salinity, the vertical salinity profile along the Wu River estuary shows the detailed changes in the salinity structure with sea level rise. Figures 10 and 11 present the distributions of tidal-averaged salinity along the Wu River estuary under the Q_{10} and Q_{90} flows to represent the high and low flow conditions, respectively, for the present condition and the sea level rise scenario. It is clear that the salinity changes throughout the entire estuary. The limit of salt intrusion is represented by a 1 psu isohaline. The limits of salt intrusion are 3000 m and 6500 m for the present condition and the sea level rise scenario under Q_{10} flow conditions (Figure 10), while they are 5500 m and 8250 m for the present condition and the sea level rise scenario under Q_{90} flow conditions (Figure 11). These two figures indicate that sea level rise pushes the limit of salt intrusion farther upstream in the Wu River estuary. The intensified stratification results in stronger gravitation circulation, which raises the salt content by transporting more saline water into the estuary. However, the sea level rise did not change the tidal amplitude, but the water surface elevation increased in the sea level rise scenario. Moreover, the sea level rise extends to the tidal excursion farther upstream, 500 m and 900 m, respectively, under the Q_{10} and Q_{90} flow conditions (not shown in the figure).

Figure 10. Distribution of the tidal-averaged salinity along the Wu River estuary under the Q_{10} flow condition for (**a**) the present condition and (**b**) the sea level rise scenario. Note that the unit of salinity is psu (practical salinity unit).

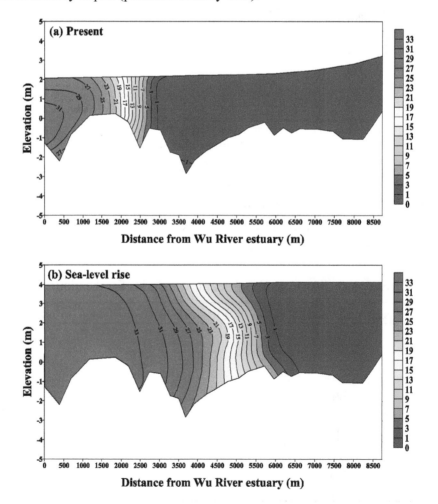

Figure 11. Distribution of the tidal-averaged salinity along the Wu River estuary under the Q_{90} flow condition for (**a**) the present condition and (**b**) the sea level rise scenario. Note that the unit of salinity is psu.

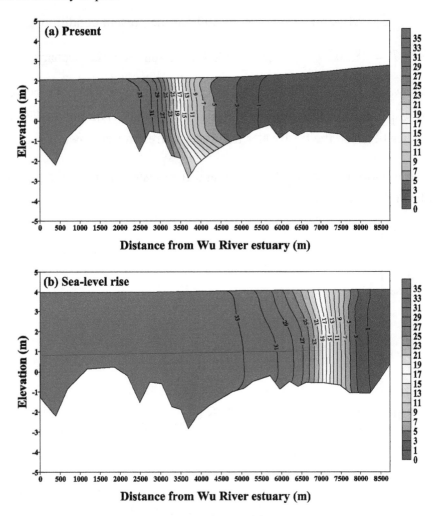

Hong and Shen [9] demonstrated that the mean salinity at the mouth and the water depth within Chesapeake Bay would increase with sea level rise. Bhuiyan and Dutta [8] also described that the sea level rise impact on salinity intrusion would be highly significant. In this study, the limit of salt intrusion will increase 3000 m and 2750 m under high and low flow conditions, respectively, for a 38.27 cm sea level rise. The maximum increased salinity reached 14.2 psu under the Q_{90} low flow. The increased salinity could cause socio-economic problems; the saline water would be unsuitable for drinking and industrial purposes. Salinity intrusion due to sea level rise would constrain the supply of water resources in the river.

5.2. Flushing Time in Response to Sea Level Rise

To calculate the flushing time in the estuary, different freshwater discharges shown in Table 2 were used to serve as the upstream boundary condition for the present condition and for the sea level rise scenario. The model simulated flushing time is plotted against river flow in Figure 12. The increase in river discharge is accompanied by a more rapid exchange of freshwater with the sea. The volume of fresh water accumulated in the estuary increases to a lesser extent compared to the volume in the discharge. Thus, the flushing time decreases with increasing river discharge. Least squares regression fitting by the power law [39] was conducted to express the empirical function for the present condition and the sea level rise scenario:

$$T_f = 1030.54 \cdot Q^{-0.821}, \ R^2 = 0.99 \text{ for the present condition} \tag{17}$$

$$T_f = 384.02 \cdot Q^{-0.599}, \ R^2 = 0.98 \text{ for the sea level rise scenario} \tag{18}$$

where Q is the freshwater discharge. With a higher correlation value (R^2), the power law statistically fits the data better, especially in the low and the high flow ends. The power law reasonably shows physical characteristics between the freshwater fraction and freshwater flow.

Figure 12. Regression between flushing time and freshwater input for the present condition and the sea level rise scenario.

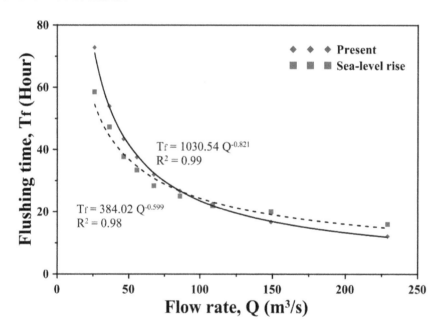

The flushing time is between 12.09 and 72.84 h under the present condition, while it is between 16.04 and 58.57 h under the sea level rise scenario. The results also indicate that the flushing time for high flow under the present condition is lower compared to the sea level rise scenario, while the flushing time for low flow under the present condition is higher compared to the sea level rise scenario. The freshwater volume thus increases under the sea level rise during high flow, and it decreases during low flow. Huang [39] applied a three-dimensional model to estimate the distributions of salinity and the freshwater fractions for flushing time estimation. He found that for the seven-day averaged flow ranging from 10 m³/s to 50 m³/s for a small estuary of North Bay, Florida, corresponding flushing time varies from 3.7 days to 1.8 days. The flushing time in the estuary of North Bay was similar to that in the Wu River estuary.

5.3. Residence Time in Response to Sea Level Rise

Passive tracers are used to simulate the material transport coming from the main river sources at the Dadu Bridge. Changing water levels and the propagation of tidal waves also result in changes in the residence time of water bodies and water constituents within the estuary and in changes in transport time through the estuary towards the sea.

The validated model was applied to explore the impact of sea level rise on residence time in the estuary. We calculated the residence time of the entire Wu River estuary under different freshwater discharge scenarios. After instantly releasing tracers throughout the entire Wu River estuary, the residence time corresponded to the time when the average tracer concentration reached its e-folding value (*i.e.*, e^{-1} value). Model results reveal that the residence time decreases as the freshwater input increases for the present condition and the sea level rise scenario (shown in Figure 13). Finding a general regression relationship between the residence time and the freshwater input would be helpful in understanding the physical and hydrological processes in the estuary. Huang *et al.* [40] conducted regression analyses between estuarine residence time and freshwater input using a power-law function in Little Manatee River, Florida. The authors found that regression by the power law provided a better fit compared to an exponential function. A regression of the residence time (T_r) *versus* freshwater input (Q) was performed and indicated an excellent correlation (R^2) through the power law function:

$$T_r = 207.18 \cdot Q^{-0.541}, \ R^2 = 0.99 \ \text{for the present condition} \tag{19}$$

$$T_r = 143.36 \cdot Q^{-0.385}, \ R^2 = 0.99 \ \text{for the sea level rise scenario} \tag{20}$$

The residence time is between 10.51 and 34.23 h under the present condition, while it is between 17.11 and 38.92 h under the sea level rise scenario. The residence time of the entire Wu River estuary increased 4.7 to 6.6 h based on different freshwater inputs due to sea level rise. The prolonged residence time will result in the deterioration of water quality and induce the limited application of water resources.

Figure 13. Regression between residence time and freshwater input for the present condition and the sea level rise scenario.

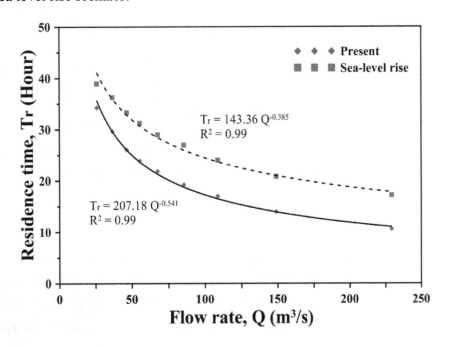

Hong and Shen [9] estimated that the residence time could increase five to 20 days in response to different sea level rise scenarios in Chesapeake Bay. The increase of residence time response to sea level rise in the Wu River estuary is smaller than that in Chesapeake Bay, because the entire estuarine system in Chesapeake Bay is much larger than the Wu River estuary.

If the sea level rise rate is changed, the salinity intrusion and transport time scales would be changed. The increase in the sea level rise rate may extend the limit of salt intrusion farther upstream and increase the residence time in the estuary. In a future study, the wind forcing and seasonal freshwater discharge input from the Dadu Bridge can be considered in the model simulations to comprehend how the salinity intrusion and transport time scales respond to these factors.

6. Conclusions

A three-dimensional hydrodynamic and salt transport model, SELFE, was established to simulate the hydrodynamics and salinity distributions in the Wu River estuary and adjacent coastal sea in northern Taiwan. The model was calibrated and verified using observational amplitudes and phases, water surface elevations and salinity distributions in 2011. The model simulation results agree well with the field observations.

The validated model was used to perform a series of numerical experiments to identify the potential impacts of future sea level rise on salinity intrusion and transport time scales, including flushing time and residence time, in the Wu River estuary of central Taiwan. The model results indicate that salinity intrusion moves farther upstream by 2750 m and 3500 m under Q_{90} and Q_{10} flow conditions, respectively, due to sea level rise. The flushing time is between 12.09 and 72.84 h for the present condition, and it is between 16.04 and 58.57 h for the sea level rise scenario. We found that the flushing time for high flow under the present condition is lower compared to the sea level rise scenario, while the flushing time for low flow under the present condition is higher compared to the sea level rise scenario. The residence time of the entire Wu River estuary increased by 23.7 h and 21.8 h for high and low flows, respectively, during the sea level rise scenario. We found that the climate change (*i.e.*, sea level rise) scenario implies not only a change in salt intrusion, but also an increase in the residence time. Sea level rise would alter the location of the river estuary, thereby causing a greater change in fish habitat and breeding ground location. Fishes breed in estuarine systems and develop in brackish waters, which is where fresh water and salt water mix. Sea level rise would move this interface backward, changing the habitat of fishing communities in the estuarine system. The increases in transport time scales (*i.e.*, residence time) due to sea level rise would prolong the transport of dissolved substances in the estuary, resulting in the deterioration of water quality.

Acknowledgments

The project under which this study was conducted is supported by the National Science Council, Taiwan, under grant no. NSC 101-2625-M-239-001. The authors would like to express their thanks to the Taiwan Water Resources Agency for providing the observational data. Appreciation and thanks are also given to the anonymous reviewers for their constructive comments and suggestions to improve this paper.

References

1. Yu, Y.F.; Yu, Y.X.; Zuo, J.C.; Wan, Z.W.; Chen, Z.Y. Effect of sea level variation on tidal characteristic values for the East China Sea. *China Ocean Eng.* **2003**, *17*, 369–382.

2. IPCC. Climate Change 2007: The Physical Science Basis. In *Contribution of Working Group 1 to the Fourth Assessment Report of the Intergovernmental Panel on Climate Change*; Solomon, S., Qin, S., Manning, M., Chen, Z., Marquis, M., Averyt, K.B., Tignor, M., Miller, H.L., Eds.; Cambridge University Press: Cambridge, UK; New York, NY, USA, 2007.

3. Hsu, H.H.; Chen, C.T. Observed and projected climate change in Taiwan. *Meteorol. Atmos. Phys.* **2002**, *79*, 87–104.

4. Yu, P.S.; Yang, T.C.; Kuo, C.C. Evaluating long-term trends in annual and seasonal precipitation in Taiwan. *Water Resour. Manag.* **2006**, *20*, 1007–1023.

5. Chiu, M.C. Relationship of Stream Insects with Flooding and Dippers in Wuling Area (in Chinese). Master Thesis, National Chung Hsing University, Taichung, Taiwan, 2009.

6. Poff, N.L.; Brinson, M.M.; Day, J.W., Jr. *Aquatic Ecosystems and Global Climate Change*; Pew Center on Global Change: Arlington, VA, USA, 2002; p. 45.

7. Hilton, T.W.; Najjar, R.G.; Zhong, L.; Li, M. Is there a signal of sea level rise in Chesapeake Bay salinity? *J. Geophys. Res.* **2008**, *113*, doi:10.1029/2007JC004247.

8. Bhuiyan, M.J.A.N.; Dutta, D. Assessing impacts of sea level rise on river salinity in the Gorai river network, Bangladesh. *Estuar. Coast. Shelf Sci.* **2012**, *96*, 219–227.

9. Hong, B.; Shen, J. Responses of estuarine salinity and transport processes to potential future sea level rise in the Chesapeake Bay. *Estuar. Coast. Shelf Sci.* **2012**, *104–105*, 33–45.

10. Lucas, L.V. Implications of Estuarine Transport for Water Quality. In *Contemporary Issues in Estuarine Physics*; Valle-Levinson, A., Ed.; Cambridge University Press: Cambridge, UK, 2010; pp. 273–306.

11. Hull, C.H.J.; Tortoriello, R. *Sea Level Trend and Salinity in the Delaware Estuary*; Staff Report; Delaware Basin Commission: West Trenton, NJ, USA, 1979.

12. Grabemann, H.; Grabemann, I.; Herbers, D.; Muller, A. Effects of a specific climate scenario on the hydrograph and transport of conservative substances in the Weser estuary, Germany: A case study. *Clim. Res.* **2001**, *18*, 77–87.

13. Chua, V.P.; Fringer, O.B.; Monismith, S.G. Influence of sea level rise on salinity in San Francisco Bay. 2011, unpublished work.

14. Rice, K.C.; Hong, B.; Shen, J. Assessment of salinity intrusion in the James and Chickahominy Rivers as a result of simulated sea level rise in Chesapeake Bay, East Coast, USA. *J. Environ. Manag.* **2012**, *111*, 61–69.

15. Chen, W.B.; Liu, W.C.; Wu, C.Y. Coupling of a one-dimensional river routing model and a three-dimensional ocean model to predict overbank flows in a complex river-ocean system. *Appl. Math. Model.* **2013**, *37*, 6163–6176.

16. Church, J.A.; White, N.J. Sea level rise from the late 19th to the early 21st century. *Surv. Geophys.* **2011**, *32*, 585–602.

17. Snay, R.; Cline, M.; Dillinger, W.; Foote, R.; Hilla, S.; Kass, W.; Ray, J.; Rohde, J.; Sella, G.; Soler, T. Using global positions system-derived crustal velocities to estimate rates of absolute sea

level change from North American tidal gauge records. *J. Geophys. Res.* **2007**, *112*, doi:10.1029/2006JB004606.

18. Nicholls, R.J.; Cazenave, A. Sea level rise and its impact on coastal zone. *Science* **2010**, *328*, 1517–1520.

19. Tseng, Y.H.; Breaker, L.C.; Cheng, T.Y. Sea level variations in the regional seas around Taiwan. *J. Oceanogr.* **2010**, *66*, 27–39.

20. Huang, C.J.; Hsu, T.W.; Wu, L.C. *Application of Tide-Gauge and Satellite Altimetry Data to Estimate Sea Level Rise*; Report to Water Resources Agency: Taipei, Taiwan, 2009.

21. Zhang, Y.L.; Baptista, A.M. SELFE: A semi-implicit Eulerian-Lagrangian finite-element model for cross-scale ocean circulation. *Ocean Model.* **2008**, *21*, 71–96.

22. Zeng, Z.; Zhao, M.; Dickinson, R.E. Intercomparison of bulk aerodynamic algorithms for the computation of sea surface fluxes using TOGA COARE and TAO data. *J. Clim.* **1998**, *11*, 2628–2644.

23. Umlauf, L.; Buchard, H. A. generic length-scale equation for geophysical turbulence models. *J. Mar. Res.* **2003**, *61*, 235–265.

24. Lauff, G.E. Lyered Sediments of Tidal Flats, Beaches, and Shelf Bottoms of the North Sea. In *American Association for the Advancement of Science Publication No. 83*; *American* Association for the Advancement of Science: Washington, DC, USA, 1967.

25. Dyer, K.R. *Estuaries: A Physical Introduction*, 2nd ed.; John Wiley: London, UK, 1977; p. 195.

26. Liu, W.C.; Hsu, M.H.; Kuo, A.Y.; Kuo, J.T. The influence of river discharge on salinity intrusion in the Tanshui Estuary, Taiwan. *J. Coast. Res.* **2001**, *17*, 544–552.

27. Huang, W.; Spaulding, M. Modelling residence-time response to freshwater input in Apalachicola Bay, Florida, USA. *Hydrol. Process.* **2002**, *16*, 3051–3064.

28. De Brye, B.; de Brauwere, A.; Gourge, O.; Delhez, E.J.M.; Deleersnijder, E. Water renewal timescales in the Scheldt Estuary. *J. Mar. Syst.* **2012**, *94*, 74–86.

29. Zimmerman, J.T.F. Mixing and flushing of tidal embayment in the western Dutch Wadden Sea. Part I: Description of salinity and calculation of mixing time scales. *Neth. J. Sea Res.* **1976**, *10*, 149–191.

30. Takeoka, H. Fundamental concepts of exchange and transport time scales in a coastal sea. *Cont. Shelf Res.* **1984**, *3*, 311–326.

31. Liu, W.C.; Chen, W.B.; Kuo, J.T.; Wu, C. Numerical determination of residual time and age in a partially mixed estuary using three-dimensional hydrodynamic model. *Cont. Shelf Res.* **2008**, *28*, 1068–1088.

32. Zhang, W.G.; Wilkin, J.L.; Schofield, O.M.E. Simulation of water age and residence time in the New York Bight. *J. Phys. Oceanogr.* **2010**, *40*, 965–982.

33. De Brauwere, A.; de Brye, B.; Blaise, S.; Deleersnijder, E. Residence time, exposure time and connectivity in the Scheldt Estuary. *J. Mar. Syst.* **2011**, *84*, 85–95.

34. Kenov, I.A.; Garcia, A.C.; Neves, R. Residence time of water in the Mondego estuary (Portugal). *Estuar. Coast. Shelf Sci.* **2012**, *106*, 13–22.

35. Shi, J.; Li, G.; Wang, P. Anthropogenic influences on the tidal prism and water exchange in Jiaozhou Bay, Qingdao, China. *J. Coast. Res.* **2011**, *27*, 57–72.

36. Hsu, M.H.; Kuo, A.Y.; Kuo, J.T.; Liu, W.C. Procedure to calibrate and verify numerical models of estuarine hydrodynamics. *J. Hydraul. Eng. ASCE* **1999**, *125*, 166–182.

37. Officer, C.B. *Physical Oceanography of Estuaries (and Associated Coastal Waters)*; Wiley: New York, NY, USA, 1976.

38. Geyer, W.R. Influence of wind on dynamics and flushing of shallow estuaries. *Estuar. Coast. Shelf Sci.* **1997**, *44*, 713–722.

39. Huang, W. Hydrodynamic modeling of flushing time in small estuary of North Bay, Florida, USA. *Estuar. Coast. Shelf Sci.* **2007**, *74*, 722–731.

40. Huang, W.; Liu, X.; Chen, X.; Flannery, M.S. Critical flow for water management in a shallow tidal river based on estuarine residence time. *Water Resour. Manag.* **2011**, *25*, 2367–2385.

3

Understanding Irrigator Bidding Behavior in Australian Water Markets in Response to Uncertainty

Alec Zuo [1], **Robert Brooks** [2], **Sarah Ann Wheeler** [1,*], **Edwyna Harris** [3] and **Henning Bjornlund** [1]

[1] School of Commerce, University of South Australia, Adelaide 5000, Australia;
   E-Mails: alec.zuo@unisa.edu.au (A.Z.); henning.bjornlund@unisa.edu.au (H.B.)

[2] Faculty of Business and Economics, Monash University, Caulfield East 3145; Australia;
   E-Mail: robert.brooks@monash.edu.au

[3] Department of Economics, Monash University, Clayton 3168, Australia;
   E-Mail: edwyna.harris@monash.edu.au

* Author to whom correspondence should be addressed; E-Mail: sarah.wheeler@unisa.edu.au;

External Editor: Miklas Scholz

Abstract: Water markets have been used by Australian irrigators as a way to reduce risk and uncertainty in times of low water allocations and rainfall. However, little is known about how irrigators' bidding trading behavior in water markets compares to other markets, nor is it known what role uncertainty and a lack of water in a variable and changing climate plays in influencing behavior. This paper studies irrigator behavior in Victorian water markets over a decade (a time period that included a severe drought). In particular, it studies the evidence for price clustering (when water bids/offers end mostly around particular numbers), a common phenomenon present in other established markets. We found that clustering in bid/offer prices in Victorian water allocation markets was influenced by uncertainty and strategic behavior. Water traders evaluate the costs and benefits of clustering and act according to their risk aversion levels. Water market buyer clustering behavior was mostly explained by increased market uncertainty (in particular, hotter and drier conditions), while seller-clustering behavior is mostly explained by strategic behavioral factors which evaluate the costs and benefits of clustering.

Keywords: water allocation market; price clustering; uncertainty; strategic behavior

1. Introduction

Water scarcity has emerged in many semi-arid regions of the world. This requires the development of mechanisms to efficiently reallocate available resources between competing extractive as well as in-stream uses. Water markets have been promoted as an efficient way of facilitating this process in a number of jurisdictions, such as Australia, USA and Chile [1–3] and more recently in Canada [4] and Spain [5]. As scarcity intensifies, demand for, and participation in, water markets is likely to increase. A continual review of market mechanisms will help to improve and facilitate greater market efficiency, through reducing transaction costs, improving product choice or reducing barriers to trade. Adoption of water market trading (where available) will represent one potential adaptation strategy for many irrigators in the face of climate change. Modeling by Adamson, Mallawaarachchi and Quiggin [6] demonstrates that adaptation will partially offset the adverse impact of climate change and suggests that improvements in the function of water markets could support adaptation.

In order to provide greater insights into how to best improve water market mechanisms and water management in general, a better understanding of irrigators' behavior in such markets and how this compares to behavior in other financial markets is necessary. In the Murray-Darling Basin (MDB) of Australia, irrigators' participation in the water market has been growing over the past two decades and this provides a unique opportunity to study irrigators' water market behavior. Two major forms of water markets exist in the MDB: the water allocation market (also known as temporary water markets, which involve the short-term right to use of water) and the water entitlement market (also known as permanent water markets involving the long-term right to access water—see Wheeler *et al.* [7] for more detail). This paper focuses on the water allocation market.

Since the Council of Australian Governments water reform agenda in 1994, water markets have played a central role in allowing farmers to deal with increased volatility, risk and adjustment pressures by permitting them to alter their short and long-term access to water resources as well as allowing them to exit out of irrigation while realizing their water assets [8,9]. In 2011, the Murray-Darling Basin Authority (MDBA) released the MDB Plan, with a target of 2750 GL to be returned from consumptive to environmental use [10]. Water entitlements are to be sourced from willing sellers, and are bought by the Commonwealth of Australia. Increasingly, there are arguments that governments should also consider buying water from the allocation market (otherwise known as temporary water available in one season) to provide environmental flows [11]. The rationale for government utilizing the water allocation market is that benefits of carry-over, lower water allocation prices, and temporal demand can provide a more efficient and flexible supply of water to meet stochastic environmental flow requirements since the timing of entitlement releases does not correspond well with the volume and timing of water applications required to achieve environmental objectives [3,12].

In light of these policy arguments regarding government intervention in the allocation market, a more thorough understanding of irrigators' trading behavior in that market, particularly how they bid and offer for water, is needed. In particular, we need to understand how variability in climate conditions impacts on water market trading behavior. One way of analyzing water market trader behavior is to analyze the extent to which irrigator bids or offers exhibit price clustering (that is, the extent they cluster around particular numbers). The existence of clustering is important as it identifies a possible dead-weight loss that exists in water allocation markets. Utilizing bid and offer data also allows us to understand how differently buyers and sellers act in the water market, something that is difficult to do in other water market analysis. It also offers continuing insights into how irrigators behave in water markets, and how similar (or dissimilar) their behavior is to participation in financial markets. Understanding the similarity between irrigator bidding and offering behavior in a water market and a trader in a stock market may also offer insights into how well introducing other water market products (for example: option trading) will be received. For instance, Heaney *et al.* [13] discuss how missing options markets in storage and delivery might impact water trading. Addressing these issues is a function of market design. Hence, undertaking analysis on price clustering is informative for water management policies aiming to improve the efficiency and flexibility of resource allocation.

Price clustering in financial markets has been well documented in the literature (e.g., Chung and Chiang [14]). Clustering is found when indicative quotes for currencies end mostly around particular numbers, for example, those with either "zero" or "five". Round numbers are disproportionately represented in bid-ask spreads for major currencies. Typically, the economics and psychological literature identify different reasons for clustering. In economics, clustering is considered a rational response to trading impediments. In psychology, clustering is thought to occur due to a human bias for prominent numbers, such as zero and one. There have been very few studies that have analyzed clustering in other non-financial product markets. There are a number of similarities between financial markets and water markets, but there are also fundamental differences because water markets are dealing with common property resource issues. In addition there is an issue as to whether recent events in financial markets during the global financial crisis make such markets an appropriate benchmark for comparison of resource markets. The issue of the efficient markets hypothesis (EMH) and the global financial crisis (GFC) is discussed by Ball [15] and Brown [16] who suggests that the failure to predict the bursting of the real estate bubble-that lead to the GFC-is in fact consistent with the central idea in the EMH. This paper analyzes clustering in the water allocation market over the past decade. In doing so, we will be able to determine (a) the extent of price clustering in this market and (b) given the constraints that prevent traders having a precise valuation of water, whether clustering behavior is a response to uncertainty (either weather or policy changes) or a strategic behavior.

The only other paper that has examined price clustering in water markets [17] found robust evidence of clustering in the water market in northern Victoria from 2002 to 2007. Its' econometric modeling suggested uncertainty faced by irrigators is a major reason for clustering. This paper extends the work by Brooks, Harris and Joymungul [17] in four ways. First, a longer time span is used covering 10 trading seasons in the northern Victorian water market. Much of this data is not publicly available. Second, alternative price clustering definitions are employed to check the robustness of the findings. Third, a variety of other data are included to identify specific factors associated with irrigators' risk awareness that in turn influence the extent of price clustering. In particular, we are interested in assessing how government water policy changes, rainfall and evaporation influence bid and offer behavior in the water market. Finally, the extent to which price clustering is a result of traders' response to uncertainty and/or strategic behavior is examined.

2. Study Area

The Goulburn Murray Irrigation District (GMID) is Australia's largest irrigation district, located in northern Victoria, along the River Murray. It has one of Australia's longest running water markets with the bulk of trading taking place in three trading zones. The most active trading zone is Greater Goulburn, which provides the data source for this paper. Water allocations and entitlements have been traded since the early 1990s and irrigators are increasingly adopting water trading (in particular water allocation trading) over time [18]. Given that the majority of trades, especially bids and offers for water, are in the water allocations market, this is the market we chose to focus on for a study on price clustering in water markets. Dairy, fruit and, grape producers are the most significant buyers in the allocation market, whereas cereal, grazing and mixed farmers are the main sellers [19]. Over the past decade, MDB irrigators have faced considerable changes to their water allocation levels (which conversely influence the amount of land irrigated). An allocation level refers to the percentage of water entitlements that is available for the entitlement holder to use throughout a season. The resource manager manages seasonal allocation levels on behalf of all entitlement holders and regularly reviews the water budget calculations in the GMID. For example, Goulburn water allocation levels dropped from a consistently secure 200% in the early 1990s to around 30% in the mid 2000s. As a result, uncertainty for irrigators has increased considerably with opening allocations of 0% in eight consecutive years from 2002 and below 100% since 1998 and with closing allocations below 100% for five out of eight years from 2002 to 2010 [18]. In 2010 and 2011 higher than expected rainfall increased water allocations, this in turn increased the amount of land irrigated. Furthermore, water policy changes add to the climate uncertainties experienced by irrigators. There have been many government and institutional changes that impact on water markets in Australia over the time-period studied. This paper considers three of the most major ones that occurred, namely: (a) the lifting of the Cap (in 1994 the Victorian Government restricted the volume of water access entitlements that could be traded out of each irrigation district in Northern Victoria to no more than 2% annually of the volume of entitlement held in the district at the start of the irrigation season. On 1 July 2006 this was increased to 4%); (b) introduction of unbundling (this occurred on 1 July 2007 in northern Victoria and it is the legal separation of rights to land and rights to access water, have water delivered, use water on land or operate water infrastructure, all of which can be traded separately) and (c) the times when the Australian Government is conducting a tender in buying back water entitlements (the Federal government began a decade long policy of buying back water entitlements from willing sellers in February 2008 in order to return water from a consumptive to an environmental use—see [7,8] for more detail).

3. Price Clustering Literature and Applications to Water Markets

3.1. Price Clustering Theories

Empirical studies in the finance literature find that the degree of clustering in any market is a function of market structure, uncertainty, resolution costs and human preferences [14]. Several hypotheses have been developed to better understand why clustering occurs. These include: the negotiation hypothesis; the price resolution hypothesis (uncertainty); the attraction hypothesis and strategic behavior. We discuss briefly each of these hypotheses and their relevance in the context of the Australian water market.

A market's structure may bring about clustering and Harris [20] developed the negotiation hypothesis to explain these effects, arguing that regulatory restrictions can reduce negotiation costs for traders. These restrictions require quotes and transaction prices to be stated as some multiple of a minimum price variation, or trading tick. Negotiation costs fall because restrictions create a discrete price set around which traders bid and offer. In the absence of these restrictions, the number of possible offers and counter-offers widens so that negotiation time also increases, creating higher price risks for participants [21]. A discrete price set reduces the amount of information exchanged, leading price to converge more quickly than would otherwise occur. As a result, transactions costs are reduced. The bid prices in the Australian water market analyzed are not required to be some multiple of a trading tick greater than one cent. Therefore, the degree of clustering is expected to be small because irrigators bid on a continuous price set.

The method of trading can also influence the degree of clustering observed. For example, the use of electronic trade compared with floor trade (in person) alters the costs associated with precise valuation and, therefore, clustering. Chung and Chiang [14] found extreme clustering occurred on floor-traded futures compared with those traded electronically. Floor trade made precise valuation more costly because it takes more time to call out information to the accuracy of several digits and there is a wider margin for error in doing so [21]. The mechanism for water trading in the GMID creates a pool price that tends to decrease the costs associated with precise valuation, so a finer grid of numbers may be expected. However, a uniform pool price each trading week may also decrease the benefits of a precise valuation and the weekly trading frequency may be too long for traders to place more precise bids. Nevertheless, Brooks, Harris and Joymungul [17] found evidence of clustering on bid prices in the GMID water market.

The price resolution hypothesis contends that prices may be evenly clustered at particular points if valuation is indecisive [22]. Loomes [23] and Butler and Loomes [24] argued that economic decision makers do not measure utilities exactly but act in a sphere of haziness, which represents the degree of difficulty in precise valuation. In other words, the risk of taking certain actions increases with uncertainty. A greater sphere of haziness implies a higher clustering propensity due to people's risk aversion behavior. When uncertainty and volatility are high, precision valuation is costly, leading to greater clustering [25].

In the case of the water market, water availability uncertainty can be brought about by several factors, including rainfall variation, water allocations, demand fluctuations, government policy changes, and climate change [7,19,26]. Variable and unpredictable rainfall in the MDB system can be on a range of time scales and intra-season variations, making it difficult to forecast final closing allocations. Allocations are announced fortnightly during the water season, and as discussed often have started at 0%. Uncertainty in allocations can lead to miscalculations regarding seasonal allocations by irrigators at the time of planting decisions. If an irrigator overestimates what their expected allocations will be at the time of planting a crop, they may have to buy additional water later. Alternatively, if an irrigator underestimates the final allocations they will receive, they may have surplus temporary water available that can be sold in the market at a later point in the season or be carried forward into the next season (depending on storage availability). Climate information only becomes available as the season progresses, so depending on how accurate irrigators were in their water expectations and the watering requirements of their permanent or annual crops, changes (or lack of changes) in monthly seasonal allocations may cause relatively high price volatility in the market. Government intervention in water markets has increased considerably over the first decade of the 21st century [27]. Government intervention affects short- and medium-term price expectations, thereby increasing costs of precise valuations.

Ikenberry and Weston [28] demonstrate that clustering of U.S. stock price also stems from the psychological preferences of market participants. This is broadly referred to as the attraction hypothesis and it suggests that clustering is the result of behavioral idiosyncrasies (heuristics). Tversky and Kahneman [29] argued that individuals often rely on a number of heuristic principles that reduce complex tasks, such as valuation to simpler or even non-optimal judgment operations.

Brooks, Harris and Joymungul [17] use variables representing the price resolution hypothesis to explain price clustering in the GMID water market. Their results indicate a large proportion of the variation in price clustering cannot be explained by the price resolution hypothesis (the largest adjusted R^2 in their regression models is 0.61). Therefore, the attraction hypothesis is very likely to be able to explain some of the variation in price clustering that cannot be explained by the models of Brooks, Harris and Joymungul [17]. Unfortunately, it is almost impossible to collect data on testable variables representing the attraction hypothesis.

An alternative explanation for clustering is that its existence is the result of strategic behavior—where people estimate the net benefits of their action [30]. Specifically, they weigh the benefits of increasing the precision of their bid/offer relative to the loss of value resulting from an imprecise estimate. In Victoria, the benefits of precise valuation are not obvious for individual traders on the water market because the water exchange Watermove used a pool price. Watermove was a trading organization in the GMID that conducted water exchanges within MDB trading zones, it operated by telephone and online. It closed down in August 2012, but still remains a valuable source of historical data, especially bid and offer data, and is used in the analysis here. Table 1 presents an example of how the Watermove exchange worked. For example, in the week of 8 September 2011, there were 35 sale offers with the offering price ranging from $14 to $100 and a total volume for sale of 4724.8 ML; and 21 buy bids ranging from $10 to $26.38 with a total volume for purchase of 5841 ML. A pool price of $21.15 (the average price of the last fulfilled sale offer, $20, and buy bid, $22.3—which is calculated after all bids and offers are received for the week) was found for the week in order to maximize the volume traded, namely 1441.5 ML. As a result, the last fulfilled buy bid had bought only 80.5 ML, instead of the full amount, 200 ML. This exchange mechanism results in the potential for price clustering to create a deadweight loss. The size of the deadweight loss depends on the pool price, the last fulfilled sale offer and buy bid prices and the amount of unsatisfied volume to sell or buy. It can be evident that the pool price could be quite different from their offer prices, which is likely to be caused by the weekly trading frequency. In this setting, the cost of rounding will be the lower likelihood of their orders being executed and the cost of not rounding will be the extra expenditure paid by buyers or the reduced revenue for sellers. A strategic bidder, therefore, would evaluate whether the cost of rounding outweighs the cost of not rounding in order to decide the bid price.

Table 1. An example of Watermove weekly exchange bids and offers.

Seller Offer Price ($/ML)	Volume for Sale	Total Volume in Exchange	Buyer Bid Price ($/ML)	Volume for Purchase	Total Volume in Exchange
14.00	200	200	26.38	200	200
15.00	103	303	25.50	200	400
15.00	18.2	321.2	25.00	400	800
15.00	60	381.2	25.00	11	811
17.00	80	461.2	25.00	200	1011
18.00	55	516.2	25.00	150	1161
19.00	320	836.2	23.38	200	1361
19.90	100	936.2	22.30	200	1561
20.00	100	1036.2	22.00	20	1581
20.00	150	1186.2	20.38	200	1781
20.00	120	1306.2	20.00	100	1881
20.00	59	1365.2	20.00	100	1981
20.00	24.3	1389.5	19.85	1500	3481
20.00	52	1441.5	18.00	500	3981
28.00	50	1491.5	18.00	200	4181
28.00	210	1701.5	15.88	200	4381
29.00	92	1793.5	15.00	200	4581
30.00	379	2172.5	15.00	10	4591
30.00	150	2322.5	14.22	500	5091
30.00	150	2472.5	12.88	500	5591
30.00	60	2532.5	10.00	250	5841
30.99	300	2832.5			
30.99	140	2972.5			
35.00	500	3472.5			
42.38	490.7	3963.2			
42.38	192.6	4155.8		Date: 8 September 2011	
45.00	68	4223.8		Pool price: $21.15/ML	
45.00	20	4243.8	Total volume traded: 1441.5 ML (The shaded bids		
45.00	100	4343.8	and offer orders were executed, with the buy order		
50.00	20	4363.8	indicated by asterisk only fulfilled by 80.5 ML)		
50.00	70	4433.8			
58.00	100	4533.8			
60.00	50	4583.8			
60.25	46	4629.8			
100.00	95	4724.8			

Traders who expect natural clustering can easily change their offer prices by a cent (penny) to avoid cluster points thereby increasing the probability of their offers being executed, described as the "pennying behavior" by Jennings [31] and also documented in Edwards and Harris [32]. This behavior is evident in the water market as demonstrated in Section 4.2. First, price clustering would decrease when traders seek a higher probability of their orders being executed. On the buyers' side, traders would require a higher probability of their orders being filled if they had overestimated seasonal allocations and therefore have experienced a deficit in available water. Assuming crop loss is a distinct possibility in this case; traders would avoid clustering to increase the likelihood that they will obtain water. On the sellers' side, greater precision could be used if surplus water could be sold at a premium price; for example, during times of protracted drought. The high returns available during these periods would encourage a greater determination for offers to be executed. Second, when buyers (sellers) consider the extra dollar expenditure (revenue) as more significant, that is, the costs of not rounding as considerable, price clustering is expected to increase. It is difficult to identify which of these effects will dominate strategic behavior in the water market, as this will depend on the market and biophysical conditions at specific times. The analysis here will investigate the effects of those conditions on the potential for strategic clustering.

3.2. Overall Water Market Clustering Hypothesis

In summary, we propose the reasons for price-clustering behavior in the water allocation market as: (1) attraction; (2) price resolution (or uncertainty); and (3) strategic behavior. Attraction suggests traders prefer certain price points to others for psychological reasons, which is discussed in Section 3.1. Price resolution proposes that traders are more likely to cluster when they perceive uncertainty in water markets is higher. We expect the following variables will be important influences on uncertainty: trading volume, water allocation price, water entitlement price, bid-ask spreads, water allocation level, climate conditions, seasonal factors, and government policy changes. Strategic behavior explanations for price clustering (where water traders will evaluate the cost and benefits of clustering, or the costs of rounding and not rounding) would also be influenced by many of these same factors. Farmers decide to trade one more unit of water allocations if the cost (revenue) from the trade is smaller (greater) than the value of the marginal product of their additional water using activities. Hence, the bid and offer prices are likely to reflect the farming enterprises and the associated risk levels for the farming enterprises if there is water scarcity. Since price clustering is measured for the whole Greater Goulburn region, we cannot consider farming enterprise variables. This question is left for future research that needs access to data across a variety of regions or access to bid and offer individual survey records (either entitlement or allocation records).

The following sections identify evidence that support the attraction hypothesis, as well as determining the extent to which price resolution and strategic behavior can explain price clustering in the water market.

4. Price Clustering Evidence in the Greater Goulburn Water Allocation Market

Before analyzing the drivers of clustering behavior in the water market, we first determined if there was evidence of clustering. We collected weekly data from Watermove on all individual buy and sell bids, including the volumes and prices of each bid for the period August 2001 to May 2011 for Greater Goulburn in Victoria—the most active trading zone. Most of these time series data are not publicly available. The data include quite a few weeks where the total number of orders is less than 20. In order to have a sufficiently large base of bids, we calculated price clustering at monthly intervals. Orders are fewer both at the start and toward the end of each season. The analysis covers ten years, and our monthly clustering series includes 100 observations, sufficient for the subsequent regression analyses. June and

July are not included as there is usually no, or very scarce, trading in those two months. For the whole dollar amount clustering series of sell offers within the 10% range of the pool price, the number of sell offers is smaller than 30 for most of the months in the 2010/2011 season. This small number of observations makes the clustering calculation unreliable. Therefore, there are 90 months instead of 100 for this clustering series.

4.1. Evidence of Price Clustering

Table 2 provides an overview of the existence of price clustering in the Greater Goulburn water allocation market. We first examined the extent of clustering at whole dollar amounts, *versus* amounts at particular cents. Over the time period being considered, 80% of all water allocation buy bids and 96% of all sell bids were placed at whole dollar amounts. Moreover, Table 2 also illustrates that if percentages are weighted by the volume associated with each order, whole dollar clustering decreases to 73% and 92% for the buy and sell orders respectively.

Table 2. Water allocation price clustering (%) at whole dollars.

Water Trade Type	All		Within 10% of Pool Price		Within 5% of Pool Price	
	Number	ML	Number	ML	Number	ML
Buy bids	79.59	72.74	78.39	70.77	78.43	71.11
Sell offers	96.47	91.69	95.92	91.08	95.84	91.20

By including orders where prices are too distant from the pool price the extent of price clustering may be biased upward because it is less costly to be precise if a price offering is likely to be far away from the pool price. As a result, it is possible for an irrigator to be acting in a greater sphere of haziness. Therefore, we calculate the clustering at whole dollar amounts again but use only those orders whose prices are within 10% of the pool price range and then only within 5% of the pool price range. As expected, Table 2 indicates the extent of price clustering decreases when orders are constrained in a narrower range around the pool price. However, the decrease appears to be small and insignificant.

Table 3 explores the extent of clustering at specific whole dollar digits of the buy and sell offers. For buy offers that are whole dollar amounts, Table 3 shows more than half of them ended in zero, while about a fifth end in five. Results are similar if the percentages are weighted by the order volumes or if only those orders within 10% or 5% of the pool price range are used.

The extent of clustering at whole dollar amounts and at specific whole dollar digits is similar to what is found for the Greater Goulburn trading zone in Brooks, Harris and Joymungul [17], where the authors use data from 2002 to 2007. Similar to Brooks, Harris and Joymungul [16], we used Chi-squared and HHI (Herfindahl-Hirschman Index) to test the significance of price clustering in our data. The results, which are available upon request, indicate the presence of significant price clustering. To further investigate price clustering over time, we present the buy and sell offer series for clustering at whole dollar amounts in Figure 1 and for clustering at the specific whole dollar digit zero in Figure 2. Figure 1 demonstrates that neither series exhibits a clear time trend but the buy offer series appears to have a greater variation over time. An augmented Dickey-Fuller unit root test indicates the absence of a unit root for both series. In Figure 2, both series appear to vary within a wider range, especially for the sell offers, compared to the results in Figure 1. The time series of clustering at the specific whole dollar digit zero also exhibits no clear time trend and does not have a unit root.

Table 3. Water allocation price clustering at whole dollar digits (%).

Whole Dollar Digits	All		Within 10% of Pool Price		Within 5% of Pool Price	
	Number	ML	Number	ML	Number	ML
Buy offers						
0	54.17	44.82	52.14	44.53	51.61	42.99
1	9.87	11.21	10.63	11.86	11.24	11.66
2	4.58	8.64	5.32	5.39	5.83	6.60
3	1.89	2.42	2.16	2.57	2.09	2.59
4	0.76	0.88	0.94	1.20	0.83	1.19
5	20.01	20.64	18.94	20.65	17.97	20.61
6	3.99	5.08	4.27	5.61	4.22	4.88
7	2.13	2.78	2.06	2.88	2.11	3.24
8	1.74	2.62	2.39	3.97	2.92	4.94
9	0.87	0.91	1.14	1.34	1.18	1.29
Sell offers						
0	71.38	59.16	67.28	56.02	67.02	55.00
1	0.48	0.84	0.53	0.86	0.49	0.74
2	0.87	1.81	0.96	1.78	1.01	1.80
3	0.66	1.32	0.87	1.36	1.05	1.78
4	1.36	2.63	1.75	2.66	1.60	2.48
5	14.52	19.76	16.54	21.00	15.86	20.31
6	0.57	1.07	0.76	1.29	0.84	1.30
7	1.04	1.67	1.16	1.68	1.27	1.77
8	3.12	4.77	3.54	4.91	4.36	6.36
9	6.01	6.96	6.62	8.44	6.50	8.46

Figure 1. Water allocation price clustering at whole dollar amounts (all offers).

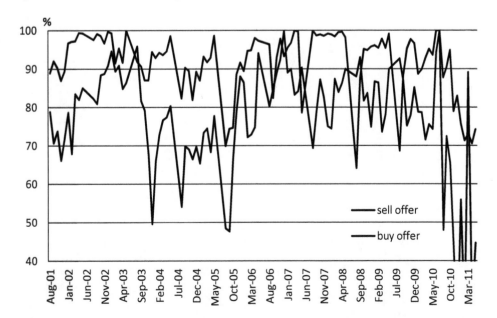

Figure 2. Water allocation price clustering at whole dollar digit ending in zero (all offers).

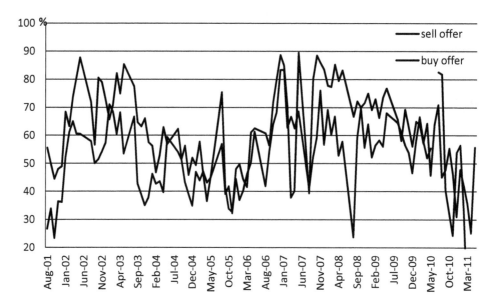

4.2. Evidence of Strategic Price Clustering Behavior

Niederhoffer [33] argues that asymmetry between ask and bid quotes around integer prices could exist because of strategic behavior where the intention is to exploit opportunities resulting from price clustering. Aşçıoğlu, Comerton-Forde and McInish [34] show that investors submit orders with one tick better than zero and five to avoid queuing orders at prices ending in these digits. Given prices cluster on round numbers, a water trader who places a bid and wants a higher probability of execution than a bid at the clustered price will tend to place the bid one cent away from the clustered price. Figures 3 and 4 investigate the evidence of strategic price-clustering behavior.

Figure 3. Distribution of buy offers not ending in whole or half dollar amounts.

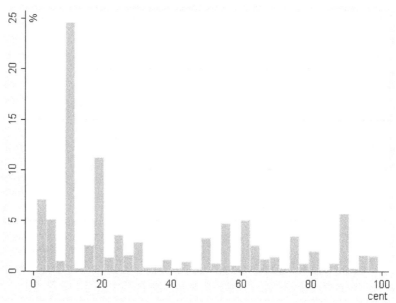

Figure 4. Distribution of sell offers not ending in whole or half dollar amounts.

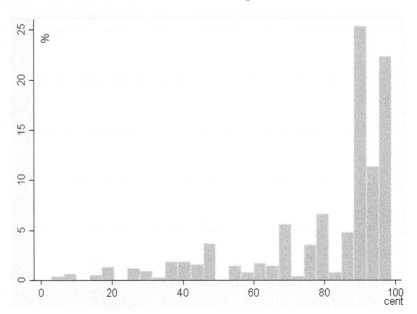

The figures respectively show the distribution of buy and sell offers that are not ended in whole or half dollar amounts. Clustering at half dollar amounts is also evident and much greater than its expected clustering. It is evident that the non-whole and half-dollar buy bids are most likely to be slightly greater than the price cluster, while the non-whole and half-dollar sell offers are mostly present slightly less than the price cluster. For those offers of whole-dollar, ending in other than zero or five, Figures 5 and 6 display the distribution across the remaining eight digits. If expecting clustering happens at zero, a buyer is most likely to place a bid with just one extra dollar. In fact, the probability of a buy bid ended in one is about 0.38-well above the probability of any other seven digits. On the sell side, a seller expecting clustering at zero is most likely to place a bid ended in nine. The probability of a sell bid ended in nine is about 0.43, well above the probability of any other seven digits.

Figure 5. Distribution of buy offers across eight digits.

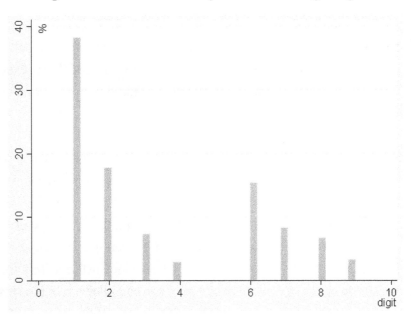

Figure 6. Distribution of sell offers across eight digits.

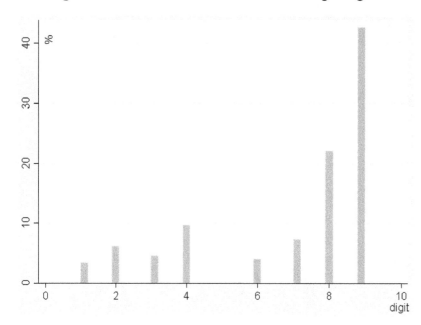

5. Methodology

Having observed substantial price clustering in the water allocation market, especially on the sell side, we now investigate the extent to which price clustering is driven by uncertainty and/or strategic behavior and if buyers and sellers' price-clustering behavior are influenced in the same way. The dependent variable, observed price clustering in a month, is defined as a proportion, which is bounded between 0 and 1. A linear probability model may not be appropriate as it can generate predictions outside the 0 and 1 interval. One way to take account of the bounded nature is the logit transformation and thus the fractional logit model, first used by Papke and Wooldridge [35]. The regression equation used was:

$$y_t^* = \log\left(\frac{y_t}{1-y_t}\right) = X_t \cdot \beta + \mu_t \tag{1}$$

where y_t is the observed price clustering in month t, X_t is a vector of regressors that potentially influence the dependent variable, and μ_t is the disturbance. The logit transformation of y_t results in a latent variable y_t^*, as a linear function of a set of regressors, X_t. The fractional logit model was executed by Stata 13's generalized linear model (GLM) command with the logit link function. We also used the type of standard error option that is heteroskedasticity- and autocorrelation-consistent to account for any heteroskedasticity and autocorrelation in the disturbance term μ_t.

y_t adopts two types of clustering weighted by order volume, namely: (1) clustering at whole dollar amounts *versus* fractions; and (2) clustering at whole dollar amounts ending in zero *versus* the remaining nine digits. For the first definition, the calculation is based on all offers and offers within the 10% range of the pool price. The clustering calculation based on offers within the 5% range of the pool price is not modeled as there is no significant difference in clustering between the 5% and 10% pool price range. For the second definition, the calculation is based on all offers since offers within the 10% range of the pool price in some months do not have enough observations to calculate a reliable clustering percentage.

Independent Variables

Table 4 lists the detailed definitions of the independent and dependent variables that were used in the price clustering models.

Table 4. Variable definitions.

Variable Name	Variable Definition
WholeBuy	Percentages of buy offers that are whole dollars in each month
WholeBuy_10	Percentage of buy offers that are whole dollars out of those within the plus and minus 10% range of pool price in each month
WholeSell	Percentages of sell offers that are whole dollars in each month
WholeSell_10	Percentage of sell offers that are whole dollars out of those within the plus and minus 10% range of pool price in each month
ZeroBuy	Percentage of buy offers that end in zero out of buy offers in whole dollars in each month
ZeroSell	Percentage of sell offers that end in zero out of sell offers in whole dollars in each month
Watervolume	Natural logarithm of volume traded for water allocations in Greater Goulburn in each month
Waterallocprice	Natural logarithm of average monthly price ($/ML) for water allocations in Greater Goulburn
Waterentprice	Natural logarithm of average monthly price ($/ML) for water entitlements in Greater Goulburn
Ln_spread	Natural logarithm of the spread between the last outstanding buyer and seller offering water allocation prices
Allocationlevel	Allocation level for Goulburn at the beginning of each month (%)
Evapminusrainfall	Monthly evaporation minus rainfall at Kerang station (mm)
Feedbarley	Natural logarithm of export price for feed barley ($/ton)
Wholemilkprice	Natural logarithm of export price for whole milk powder ($/kg)
Cattleprice	Natural logarithm of export price for cattle ($cent/kg)
Carryover %	Percentage of water entitlement allowed for carryover (note for 2010/11 season all the allocation in linked Allocation Bank Account on 30 June 2011 is eligible for carryover—there is no maximum)
Govpolicy	1 for the months when major water market policies were introduced/ongoing in the GMID (namely the lifting of the Cap, introduction of unbundling and the times when the Government is conducting a tender in buying back water). For Cap and unbundling introduction, the dummy is coded for the first three months after policy introduction
Govpolicy10/11	Interaction variable between Govpolicy variable and season 2010/11
Monthindex	Monthly index from 1 to 10 for August to May, respectively
Monthindexsqrd	Monthly index squared

Note: The first six variables are the respective dependent variables for the six regression models presented in Table 5.

Our final choice of independent variables was influenced by other studies that have studied influences on water market trade (e.g., Wheeler *et al.* [19] and Brooks *et al.* [17]). It was also determined by statistical issues, such as serious multicollinearity (discussed in Section 6). There are a number of potential relationships our independent variables could have with price clustering, and these impacts will vary depending on whether we are looking at buyer or seller behavior. For example, weather, measured by net evaporation in millimeters, may be positively related to price clustering according to the price resolution (uncertainty) hypothesis. Net evaporation is calculated as total evaporation minus total rainfall for the month in question. *Ceteris paribus*, drier weather increases water prices and, in turn, increases the uncertainty perceived by irrigators who are trying to buy water, resulting in a higher level of price-clustering behavior in the water market. However, drier weather presents a greater need for water in general. In turn, at the margin some buyers will have a greater need to have their orders executed, and therefore act more strategically in the market. This will reduce price clustering overall. Alternatively, as water prices increase sellers' risk decreases so there is less need for strategic behavior to sell their water. As a result, the overall effect of weather on price clustering depends on whether water buyers or sellers are behaving more risk aversely or strategically. Other independent variables that may be influenced by the price resolution (uncertainty) hypothesis for both buyers and sellers include water allocation and water entitlement prices, trading volume, the spread between the offer prices, feed barley prices, carryover level and government policy. Our government policy variable represents either (a) a time of uncertainty, namely three months after major policy changes, such as unbundling of land and water and the changing water trade restriction policies; and/or (b) a time when the government is purchasing water entitlements in the market. Victoria has had annual restrictions on the amount of entitlement trade allowed out of a district for years. In January 2006, the cap on entitlement trade was eased from 2% to 4%. The unbundling of land and water occurred in the GMID on 1 July 2007. Unbundling reduced the transaction costs associated with trading water, and allowed irrigators to own shares in different rivers (reducing risks). The unbundling aimed to facilitate trading in water entitlement and allocation and make trading more efficient.

Variables that may be influenced primarily by the strategic behavior hypothesis include whole-milk powder prices, cattle prices and water allocations received by irrigators, but risk averse behavior may also play a part in influencing price clustering. Whole-milk powder represents a production output of dairy farmers, feed barley represents an input substitute for watering pasture for dairy production, and cattle represents an alternative output production substitute. The overall influence of each variable will be determined by the strength of each hypothesis in determining behavior. Wherever model statistics allow, we have included all the same independent variables in every model to examine whether there are any differences between the influences on buying and selling clustering behavior.

6. Results and Discussion

Results for our buy and sell price clustering models in the Greater Goulburn water allocation market are presented in Table 5. Since the coefficient results produced by the fractional logit model are not practically meaningful, we report the marginal effect estimates. Multicollinearity was an issue in some of the models, with the variance inflation factors (VIFs) of water allocation price, water entitlement price, spread, allocation level and government policy variables being greater than five. The potential consequence is to make the variables involved insignificant where they should be significant. In order to verify whether collinearity caused this problem, we dropped the variables with insignificant coefficients one by one and checked whether the coefficients of the remaining variables became significant. If this was the case, the involved insignificant variables were dropped. However, if it was not the case they were kept in order to minimize omitted variable bias.

Table 5. Buy and sell offer monthly water allocation price clustering.

Variable	WholeBuy	WholeBuy_10	ZeroBuy	WholeSell	WholeSell_10	ZeroSell
Watervolume	−0.003	−0.034	−0.035 ***	0.016 ***	0.026	−0.032 *
Waterallocprice	0.016	0.033	-	−0.005	-	0.045
Waterentprice	0.067	0.079	-	−0.165 ***	-	-
Ln_spread	−0.015	−0.096 ***	0.050 ***	0.046 ***	-	0.063 ***
Allocationlevel	0.001	0.002	-	−0.0002	-	-
Evapminusrainfall	0.001 ***	0.001 ***	0.0003	0.00004	−0.0001	0.00002
Feedbarley	0.240 ***	0.377 ***	0.090	−0.010	0.052	−0.020
Wholemilkprice	−0.122	−0.225 *	−0.130 **	0.094 ***	0.026	0.179 **
Cattleprice	−0.059	0.011	−0.039	−0.363 ***	−0.478 ***	−0.389 *
Monthindex	−0.055 **	−0.128 ***	0.016 ***	0.009 ***	-	0.015 ***
Monthindexsqrd	0.005 **	0.011 ***	-	-	-	-
Carryover	−0.001	0.000	0.0002	−0.0003	−0.001	−0.001 **
Govpolicy	0.076 *	0.135 **	0.008	−0.028	−0.070 **	−0.090
Govpolicy10/11	−0.255 ***	−0.248 ***	−0.277 ***	−0.029	-	0.075
Observations	100	94	100	100	90	100
Log likelihood	−35.80	−32.66	−44.52	−19.34	−18.89	−42.65
BIC	−386.85	−356.94	−405.79	−393.45	−360.68	−401.31

Note: Marginal effects are reported. * $p < 0.1$; ** $p < 0.05$; *** $p < 0.01$ indicate significance at the 10%, 5% and 1% levels, respectively.

6.1. Buy Offer Price Clustering

Positive coefficients for net evaporation, feed barley price and the government policy dummy suggest that uncertainty (from the price resolution hypothesis) is able to explain clustering by buyers in the water market. Higher net evaporation loss increases water uncertainty and increases clustering. Higher feed barley prices augment water demand because it is an input substitute for on-farm feed production. In turn, as feed barley prices rise dairy farmers will find it more costly to replace water to grow their own pasture with purchased feed. This increases water market demand, and the costs of precise bids thereby causing greater clustering in buy offers.

The government policy dummy represented periods of uncertainty and significant government intervention in the market (e.g., the first three months following significant government changes) and is associated with greater uncertainty in water prices; especially in the short-term after the policy introduction. For two of three buy models; periods of policy uncertainty were positively and significantly associated with price clustering. This implies that water allocation buyers are using price clustering as a response to policies that add to market uncertainty. In times of change; irrigators will be operating in a greater sphere of haziness; with higher levels of uncertainty and volatility being experienced; so buyers exhibit a higher clustering propensity.

A surprising finding regarding the government policy variable is the result of its interaction with season 2010/11, when water was plentiful due to the record rainfall during the season, when prices dropped accordingly. Contrary to the positive impact of government policy on price clustering observed for previous seasons, government policy had a significantly negative impact in 2010/2011. Two influences (government intervention and rainfall) may explain this result. The Commonwealth was in the market buying entitlement water from November 2010 to May 2011, which was a time of flooding and falling water prices. The flooding reduced irrigator buyers' risk and their water demand, thereby reducing their clustering.

The price resolution hypothesis also predicts that the trade volume is negatively associated with price clustering, while price is positively related to price clustering. Our results, however, only offer a very weak support for this. The volume of trade has a significantly negative impact on clustering in the zero buy model, while a negative but insignificant impact on clustering at whole dollar. Neither water allocation nor entitlement prices have significant impact on clustering although their impacts are estimated as positive.

The coefficients of our time variable—months in the year (and its squared term)—suggests buyer price clustering generally decreases from the start of the season (August) until the month of January and then increases until the end of the season. Brennan [27] argued irrigators are generally risk averse and will hold more water than required at the start of a season when climate and allocation information is yet to be revealed, creating price premiums. As a result, some buyers may be more concerned with having their orders executed, increasing the costs of rounding. If, in the aggregate, all buyers behave this way, clustering will fall over the season. This result could also be explained by buyers' aversion to the sequential resolution of uncertainty suggesting a preference for uncertainty to be resolved all at one time rather than sequentially [36]. Hence, facing limited and uncertain climate information at the beginning of the season, buyers intend to secure the water they need at one time rather than through multiple orders as the season progresses. Later in the season (e.g., January onwards in our results) when climate and allocation information are revealed, uncertainty will diminish, the costs of rounding will decrease and therefore, clustering will increase again. The results presented in Table 5 demonstrate this outcome for most of the buy models, whereas in the sell models the opposite is true: clustering tends to increase throughout the water season.

The results that suggest strategic behavior as a reason for clustering by water buyers include the negative coefficient for whole milk powder price. When the milk powder price increases, irrigators have greater incentive to produce milk to take advantage of the higher returns. In turn, they are more determined to have their buy offers executed, so the costs of rounding increases thereby decreasing price clustering. We would expect to see the opposite effect on clustering if the price resolution hypothesis applied in this case.

But overall, it appears that buyer bid behavior in water markets is most influenced by price resolution (uncertainty) rather than strategic behavior. In light of the fact that our data-set includes years during which irrigators were learning how to use the new water market, it is not surprising that, on balance, uncertainty would create costs associated with precision thereby leading to greater clustering. The continuing tendency for clustering on the buyers' side of the market may well reflect the ongoing uncertainty caused by the combined effects of Australia's highly variable climate and changes in government policy.

6.2. Sell Offers Price Clustering

Both the price resolution hypothesis and strategic behavior can also be identified from significant variables in our seller price clustering models. The results for volume could support either hypothesis with positive coefficients in the WholeSell models and negative coefficients for the ZeroSell model. An increase in clustering in the whole sell models reflects strategic behavior where the costs of rounding are low because sellers may be less determined to have their trades executed. The price resolution hypothesis better explains the decrease in clustering in the ZeroSell model because greater trade intensity creates higher liquidity levels and produces more information with regard to value, allowing for greater precision. In combination, these factors reduce volatility and clustering. Alternatively, these mixed signs could suggest that the attraction hypothesis better explains the effects of volume on clustering for water sellers and that these traders are simply drawn to particular numbers.

The positive significant coefficients for spread lend support to the price resolution hypothesis because a wider bid-ask spread indicates precise valuations are more difficult. This adds to market volatility, so clustering will increase. The negative and significant coefficient on cattle prices is also consistent with the price resolution hypothesis. An increase in cattle prices (which is a dryland output substitute for irrigated production) would lead to a reduction in water demand and price. Falling water prices increase the costs of rounding thereby causing the clustering levels to fall also.

Water entitlement price is significantly negative in the WholeSell model, which suggests that strategic behavior, rather than the price resolution hypothesis, explains price clustering at whole dollars. When some buyers replace water entitlements with water allocations due to increasing water entitlement prices, the demand for water allocations increases and this pushes up water allocation prices. Water allocation sellers may consider the loss in revenue from pennying behavior is compensated by the higher allocation price and a greater chance of offer execution. Hence price clustering decreases and pennying behavior increases.

A positive impact of whole milk powder price, or a negative impact of carry-over level on price clustering, would suggest that strategic behavior may be playing a role in seller behavior. Our results support these hypotheses. Whole milk powder price has a positive estimate in all sell models and significant in the WholeSell and ZeroSell models, while carry-over level has a negative estimate in all sell models but is only significant in the ZeroSell model. As whole milk powder price rises, demand for water also increases so that higher returns from selling water accrue and sellers may expect to trade a higher volume. This magnifies the extra dollar per megaliter from clustering at whole dollars ending in zero, indicating strategic behavior may be utilized by sellers in these situations.

A higher carry-over percentage potentially increases the demand for water allocations in the market, especially later in the season, as risk-averse farmers can carry-over water that they have not used, and buy extra supplies to cover potential shortfalls the following season. This is a more dynamic explanation of the impact of carry-over in the water market, where irrigators are adjusting their practices over seasons. Water allocation prices are therefore higher than otherwise and price clustering decreases.

The government policy variable had a significant negative impact on price clustering in the WholeSell_10 model. In general, periods of policy uncertainty decrease price-clustering behavior by sellers, indicating that perhaps price increases are expected, there is a lower risk of entering the market for sellers and hence price-clustering behavior falls.

The relationship between most of the variables and clustering outcomes on the sellers' side of the market runs in the opposite direction to that which would be expected under the price resolution hypothesis. Therefore, it appears that strategic behavior influences seller bid behavior more than buyer bid behavior.

7. Conclusions

This paper has provided evidence to show there are a range of influences impacting buyer and sellers' water allocation market behavior in the Greater Goulburn trading zone in Victoria. While there are similarities between irrigators' behavior in the water market and general investors' behavior in the financial product markets, such as strong evidence of price clustering present in both markets, differences between two markets exist in terms of the explanations for price clustering, which we have investigated in the current study. Understanding irrigators' water market clustering behavior allows us to gain a range of possible insights about how buyers and sellers may respond to uncertainty and policy changes in the market. These insights are useful for achieving more efficient resource allocation. Our analyses indicate that buyer-clustering behavior is for the most part explained by the price resolution hypothesis—where uncertainty tends to increase risks and decrease the costs of rounding. The cost of precision valuation increases when water allocation prices are difficult to predict and are volatile. For buyers, times of severe climate conditions (e.g., hotter and drier conditions), commodity price volatility, and government policy introduction increases the risk associated with trading and, thereby, their price-clustering behavior.

Conversely, the models' results seem to reflect that sellers' clustering behavior is more reflective of strategic behavior than uncertainty. Strategic behavior in water markets prevails when the benefit of clustering does not outweigh its cost. These costs may include a reduction in the chance of order execution; an increase in the purchase cost for buyers; or an associated loss of sale revenue for water sellers. Correspondingly, the cost of unsuccessful sale offers is high if buyers are in greater need of water or if sellers keenly anticipate the revenue from water sales. Under such circumstances of high costs, traders are likely to consider carefully the cost of clustering and bid/offer strategically, which our results suggested happened the most in the seller clustering models. Hence, our results suggest sellers are acting in a more sophisticated manner in water markets than water buyers, and most of the costs of clustering are therefore borne by buyers.

In terms of policy implications from this research, it is clear that there is a need, wherever possible, for governments to attempt to reduce irrigator uncertainty. This will be of most importance for buyers. More effective farmer adaptation to external impacts, such as water variability is driven by timely and useful information. Water price, climate, commodity forecasts, allocation information and certainty in government policy are all important influences of water market strategies. Incomplete and fragmented information, as well as uncertain policy, decreases farmers' ability to manage their water needs.

Acknowledgments

This research was supported by an Australian Research Council Discovery Project DP140103946.

Author Contributions

Alec Zuo conducted the majority of the analysis, and wrote the paper with Sarah Wheeler. Robert Brooks and Edwyna Harris provided the original idea for the paper, and Henning Bjornlund provided historical water market data.

References

1. Bjornlund, H.; McKay, J. Aspects of water markets for developing countries—Experiences from Australia, Chile and the US. *Environ. Dev. Econ.* **2002**, *7*, 767–793.

2. Grafton, R.Q.; Libecap, G.; McGlennon, S.; Landry, C.; O'Brien, B. An integrated assessment of water markets: A cross-country comparison. *Rev. Environ. Econ. Policy* **2011**, *5*, 219–239.

3. Wheeler, S.; Garrick, D.; Loch, A.; Bjornlund, H. Evaluating water market products to acquire water for the environment in Australia. *Land Use Policy* **2013**, *30*, 427–436.

4. Nicol, L.; Klein, K.; Bjornlund, H. Permanent transfers of water rights: A study of the southern Alberta market. *Prairie Forum* **2008**, *33*, 341–356.

5. Giannoccaro, G.; Pedraza, V.; Berbel, J. Analysis of stakeholders' attitudes towards water markets in Southern Spain. *Water* **2013**, *5*, 1517–1532.

6. Adamson, D.; Mallawaarachchi, T.; Quiggin, J. Declining inflows and more frequent droughts in the Murray-Darling Basin: Climate change, impacts and adaptation. *Aust. J. Agric. Resour. Econ.* **2009**, *53*, 345–366.

7. Wheeler, S.; Loch, A.; Zuo, A.; Bjornlund, H. Reviewing the adoption and impact of water markets in the Murray-Darling Basin, Australia. *J. Hydrol.* **2014**, *518*, 28–41.

8. Crase, L.; Pagan, P.; Dollery, B. Water markets as a vehicle for reforming water resource allocation in the Murray-Darling Basin. *Water Resour. Res.* **2004**, *40*, 1–10.

9. National Water Commission. *Impacts of Water Trading in the Southern Murray—Darling Basin Between 2006–07 and 2010–11*; Commonwealth of Australia: Canberra, Australia, 2012.

10. Murray-Darling Basin Authority (MDBA). *Proposed Basin Plan*; MDBA: Canberra, Australia, 2011.

11. *Market Mechanisms for Recovering Water in the Murray-Darling Basin*; Final Report for Productivity Commission: Canberra, Australia, 2010.

12. Loch, A.; Bjornlund, H.; Wheeler, S.; Connor, J. Trading in allocation water in Australia: A qualitative understanding of irrigator motives and behavior. *Aust. J. Agric. Resour. Econ.* **2012**, *56*, 42–60.

13. Heaney, A.; Dwyer, G.; Beare, S.; Peterson, D.; Pechey, L. Third-party effects of water trading and potential policy responses. *Aust. J. Agric. Resour. Econ.* **2006**, *50*, 277–293.

14. Chung, H.; Chiang, S. Price clustering in E-mini and floor-traded index futures. *J. Futur. Mark.* **2006**, *26*, 269–295.

15. Ball, R. The global financial crisis and the efficient markets hypothesis: What have we learned? *J. Appl. Corp. Financ.* **2009**, *21*, 8–16.

16. Brown, S. The efficient markets hypothesis: The demise of the demon of chance? *Account. Financ.* **2011**, *51*, 79–95.

17. Brooks, R.; Harris, E.; Joymungul, Y. Price clustering in Australian water markets. *Appl. Econ.* **2013**, *45*, 677–685.

18. Bjornlund, H.; Wheeler, S.; Rossini, P. Water Markets and Their Environmental, Social and Economic Impact in Australia. In *Water Trading and Global Water Scarcity: International Perspectives*; Maestu, J., Ed.; Francis Taylor: Gloucester, UK, 2013; pp. 68–93.

19. Wheeler, S.; Bjornlund, H.; Shanahan, M.; Zuo, A. Price elasticity of allocations water demand in the Goulburn-Murray irrigation district of Victoria, Australia. *Aust. J. Agric. Resour. Econ.* **2008**, *52*, 37–55.

20. Harris, L. Stock price clustering and discreteness. *Rev. Financ. Stud.* **1991**, *4*, 389–415.

21. Grossman, S.; Miller, M.; Cone, K.; Fischel, D.; Ross, D. Clustering and competition in asset markets. *J. Law Econ.* **1997**, *40*, 23–60.

22. Ball, C.; Torous, W.; Tschoegl, A. The degree of price resolution: The case of the gold market. *J. Futur. Mark.* **1985**, *5*, 29–43.

23. Loomes, G. Different experimental procedures for obtaining valuations of risky actions: Implications for utility theory. *Theory Decis.* **1988**, *25*, 1–23.

24. Butler, D.; Loomes, G. Decision difficulty and imprecise preferences. *Acta Psycholog.* **1988**, *68*, 183–196.

25. Capelle-Blanchard, G.; Chaudhury, M. Price clustering in the CAC 40 index options market. *Appl. Financ. Econ.* **2007**, *17*, 1201–1210.

26. Mallawaarachchi, T.; McClintock, A.; Adamson, D.; Quiggin, J. Investment as an Adaptation Response to Water Scarcity. In *Water Policy Reform: Lessons in Sustainability from the Murray-Darling Basin*; Quiggin, J., Mallawaarachchi, T., Chambers, S., Eds.; Edward Elgar: Cheltenham, UK, 2012; pp. 101–126.

27. Brennan, D. Water policy reform in Australia: Lessons from the Victorian seasonal water market. *Aust. J. Agric. Resour. Econ.* **2006**, *50*, 403–423.

28. Ikenberry, D.L.; Weston, J.P. Clustering in US stock prices after decimalisation. *Eur. Financ. Manag.* **2008**, *14*, 30–54.

29. Tversky, A.; Kahneman, D. Judgement under uncertainty: Heuristics and biases. *Science* **1974**, *185*, 1124–1131.

30. Mitchell, J. Clustering and psychological barriers: The importance of numbers. *J. Futur. Mark.* **2001**, *21*, 395–428.

31. Jennings, R. Getting "pennied": The effect of decimalization on traders' willingness to lean on the limit order book at the New York Stock Exchange. *NNYSE Doc.* **2001**, *1*, 1–24.

32. Edwards, A.; Harris, J. *Stepping Ahead of the Book*; Securities and Exchange Commission: Washington, DC, USA, 2002.

33. Niederhoffer, V. Clustering of stock prices. *Oper. Res.* **1965**, *13*, 258–265.

34. Aşçıoğlu, A.; Comerton-Forde, C.; McInish, T.H. Price clustering on the Tokyo Stock Exchange. *Financ. Rev.* **2007**, *42*, 289–301.

35. Papke, L.E.; Wooldridge, J.M. Econometric methods for fractional response variables with an application to 401(K) plan participation rates. *J. Appl. Econ.* **1996**, *11*, 619–632.

36. Palacios-Huerta, I. The aversion to sequential resolution of uncertainty. *J. Risk Uncertain.* **1999**, *18*, 249–269.

4

Vulnerability Assessment of Environmental and Climate Change Impacts on Water Resources in Al Jabal Al Akhdar, Sultanate of Oman

Mohammed Saif Al-Kalbani [1,*]**, Martin F. Price** [1]**, Asma Abahussain** [2]**, Mushtaque Ahmed** [3] **and Timothy O'Higgins** [4]

[1]   Centre for Mountain Studies, Perth College, University of the Highlands and Islands, Crieff Road, Perth PH1 2NX, UK; E-Mail: Martin.Price.perth@uhi.ac.uk

[2]   Department of Natural Resources and Environment, College of Graduate Studies, Arabian Gulf University, Manama 26671, Bahrain; E-Mail: asma@agu.edu.bh

[3]   Department of Soils, Water & Agricultural Engineering, College of Agricultural and Marine Sciences, Sultan Qaboos University, PO Box 34, Al-Khod 123, Oman; E-Mail: ahmedm@squ.edu.om

[4]   Scottish Association for Marine Sciences, University of the Highlands and Islands, Scottish Marine Institute, Dunstaffnage, Argyll PA37 1QA, UK; E-Mail: Tim.o'higgins@sams.ac.uk

External Editor: Simon Beecham

*   Author to whom correspondence should be addressed;
    E-Mail: Mohammed.Al-kalbani@perth.uhi.ac.uk or Kmohd2020@yahoo.com;

Abstract: Climate change and its consequences present one of the most important threats to water resources systems which are vulnerable to such changes due to their limited adaptive capacity. Water resources in arid mountain regions, such as Al Jabal Al Akhdar; northern Sultanate of Oman, are vulnerable to the potential adverse impacts of environmental and climate change. Besides climatic change, current demographic trends, economic development and related land use changes are exerting pressures and have direct impacts on increasing demands for water resources and their vulnerability. In this study, vulnerability assessment was carried out using guidelines prepared by United Nations Environment Programme (UNEP) and Peking University to evaluate four components of the water resource system: water resources stress, water development pressure, ecological health, and management capacity. The calculated vulnerability index (VI) was high, indicating that the water resources are experiencing levels of stress. Ecosystem deterioration was the dominant parameter and management capacity was the dominant category driving the vulnerability on water resources. The vulnerability assessment will support policy and decision makers in evaluating options to modify existing policies. It will also help in developing long-term strategic plans for climate change mitigation and adaptation measures and implement effective policies for sustainable water resources management, and therefore the sustenance of human wellbeing in the region.

Keywords: vulnerability index; water resources stress; development pressure; ecological health; management capacity; climate change; adaptation; mitigation; Al Jabal Al Akhdar; Oman

1. Introduction

Freshwater resources are key ecosystem services which sustain life and all social and economic processes. Their disruption threatens the health of ecological systems, people's livelihoods and general human wellbeing. However, water resources are being degraded as a result of multiple interacting pressures [1], particularly environmental and climate changes. The Fourth and Fifth Assessment Reports of the Intergovernmental Panel on Climate Change (IPCC) played a major role in framing understanding of likely impacts of climate change on human society and natural systems, making it clear that "water is in the eye of the climate management storm" [2–4]. Different possible threats resulting from anthropogenic climate change include temperature increases, shifts of climate zones, sea level rise, droughts, floods, and other extreme weather events [5]. The Earth's surface temperature has increased by about 0.5 °C during the last two decades, and a rise with similar amplitude is expected up to 2025, with direct effects on the global hydrological cycle, impacting water availability and demand [2–4]. Negative impacts on water availability and on the health of freshwater ecosystems will have negative consequences for social and ecological systems and their processes [6]. For example, with an approximately 2 °C global-mean temperature rise, around 59% of the world's population would be exposed to irrigation water shortage [7].

Besides climate change impacts, other drivers of environmental changes such as demographic trends, economic development and urbanization and related land-use changes are exerting pressures and increase demand for water resources [8]. Together, these drivers are stressing water resources far beyond the changes caused by natural global climatic changes in the recent evolutionary past. As a result of rapid population growth and economic development, and mismanagement of water resources, these drivers exert pressures on water resources, changing them both spatially and temporally and causing imbalances between supply and demand in hydrological systems [9]. The net effects can be translated into increases in the vulnerability of water resources systems. These systems are especially vulnerable to such changes because of their limited adaptive capacity, which can create major challenges for future management of water resources for human and ecosystem needs [10]. Therefore, there is a need to assess the vulnerability of water resources in order to enhance management capacity and adapt measures to cope with these changes for sustainable water resources use and management.

Vulnerability is a term commonly used to describe a weakness or flaw in a system; its susceptibility to a specific threat and harmful event. There have been many efforts to use the concept across different fields which are often location or sector specific. A variety of definitions of vulnerability have been proposed in the climate change literature, e.g., [11–18]. Common to most is the concept that vulnerability is a function of the exposure and sensitivity of a system to a climate hazard, and the ability to adapt to the effects of the hazard [15,19]. From a social point of view, vulnerability is defined as the exposure of individuals or collective groups to livelihood stress as a result of the impacts of such environmental change or climate extremes [17,20]. In this context, vulnerability can be explained by a combination of social factors and environmental risk, where risk derives from physical aspects of climate-related hazards exogenous to the social system [21–24]. Vulnerability to climate change is generally understood to be a function of a range of biophysical and socioeconomic factors. It is considered a function of wealth, technology, education, information, skills, infrastructure, access to resources, and stability and management capabilities [14,25]. The IPCC has defined vulnerability to climate change as the degree to which a system is susceptible to, and unable to cope with, adverse effects of climate change, including climate variability and extremes [26,27]. Vulnerability is also a function of the character, magnitude, and rate of climate change and variation to which a system is exposed, its sensitivity, and its adaptive capacity [28,29]. It is widely seen as an integrative concept that can link the social and biophysical dimensions of environmental change [30,31].

Nevertheless, "vulnerability" means different things to different researchers. From a water resources perspective, vulnerability has been defined as "the characteristics of water resources system's weakness and flaws that make the system difficult to be functional in the face of socioeconomic and environmental change" [10] (p. 2). Thus, water resources vulnerability assessment is an investigative and analytical process to evaluate the sensitivity of a water system to potential threats and identify challenges in mitigating the risks associated with negative impacts, in order to support water resources conservation and management under climate and environmental changes [10]. It is a tool to identify potential risks, helping to analyse specific aspects that contribute to overall risk. It therefore provides useful information to the manager about which components should receive more focus, in order to improve water management capacity towards sustainability in adapting to the changing climate and environmental factors.

Most water-stressed arid countries are vulnerable to the potential adverse impacts of climate change; particularly increases in temperatures, less and more erratic precipitation, drought and desertification. This is especially true in arid mountain regions, particularly Al Jabal Al Akhdar where a unique set of water management practices has enabled the development and survival, over centuries, of an agro-pastoral oasis social-ecological system. This study was conducted to assess the environmental and climate change impacts on water resources of Al Jabal Al Akhdar since no vulnerability assessments have been previously conducted in Oman or in this fragile mountain ecosystem.

The overall aim of this study was to estimate the vulnerability index of water resources of Al Jabal Al Akhdar to climate and environmental changes, to establish to what extent these resources are vulnerable and identify the major risks and levels of stress it faces with regard to water stress, development pressure, ecological health and management capacity. These are essential components for computing vulnerability index and assessing water resources in the region. The results should provide decision-makers with options to evaluate the current situation, modify existing policies, and implement adaptation and mitigation measures for sustainable water resources management in the study area.

2. The Study Area

Al Jabal Al Akhdar (Green Mountain) is located in the central part of the northern western Al-Hajar Mountains of the Sultanate of Oman (Figure 1), in the highest portion of 1500 to 3000 m above sea level [32]. It is a long-established agro-pastoral oasis ecosystem which has supported communities for centuries [32]. Until the late 20th century, this social ecological system has been geographically isolated and, compared to many places, relatively closed to the outside world. Historically, water availability has connected the agro-pastoral system and dictated the bounds of agricultural development and the human development. The area is also of particular cultural significance for Omani people for its location, topography, agricultural terraces, biodiversity and climate.

Figure 1. Oman map showing the location of Al Jabal Al Akhdar (in blue rectangle).

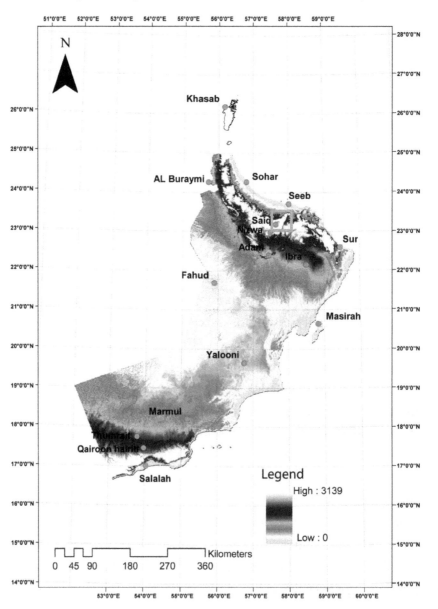

Al Jabal Al Akhdar has a Mediterranean climate. Because of its altitude, temperatures are some 10 to 12 °C lower than in the coastal plains. In general, mean monthly temperatures drop during winter to below 0 °C and rise in summer to around 22 °C. Temperature records from Saiq Meteorology Station (the only one in the area, part of Oman's national climate monitoring network), from 1979 to 2012 show mean monthly air temperatures from February to April from 12.1 to 18.7 °C and around 25 °C during summer (July and August). Minimum temperatures ranged from −0.6 °C in January to 15.9 °C in July, and maximum temperatures from 20.3 °C in January to 33.5 °C in July [33]. Rainfall is highly variable and irregular with an annual mean of about 250–400 mm [33] and is the main source of fresh water. There are two distinct rainfall seasons: the winter season from mid-November through March, and the summer season from mid-June through mid-September. From 1979 to 2012, the average annual rainfall was 295.3 mm, with the highest monthly averages of 45.8 and 42 mm during August and July; and the lowest of 8.8 and 8.2 mm during October and November, respectively [33].

Agriculture is the main economic activity, providing the basis of livelihoods for around 70% of the inhabitants [34]. Although the sector does not contribute much to the national economy (only 3.7% of the total Oman GDP), it is the main dominant water consumer in Oman including the study area (more than 92% of the total available water) [35]. The area produces a variety of fruits, particularly pomegranates and roses (grown for the extraction of rose water), which are sold in the local markets as the major sources of income for farmers. Annual crops such as garlic, onion, maize, barley, oats and alfalfa are sometimes planted in the terraces depending on the availability of water. The area also has several endangered or vulnerable species of flora and fauna that are not found elsewhere in Oman [36].

Livestock husbandry is also an important part of the agriculture in the area, for food and income through the sale of fibre (goat and sheep hair), and provides a source of manure for the cultivation of crops. Goats are the main livestock in local communities, representing more than 80% of the total animal units [37].

Due to its relatively cool weather, especially during summer, and the construction of asphalted access road up to the mountain in 2006, followed by the construction of hotels, many tourists visit the area, mainly to see natural landscapes and agricultural terraces and to camp. There has been an increase in the number of hotels in the study area, from one in 2006 to four in 2014, plus other holiday apartments and rest houses. The number of tourists has increased by 58% from 85,000 in 2006 to 134,000 in 2013 [37].

Natural freshwater resources in Al Jabal Al Akhdar are of three types: groundwater (wells), lotic resources (natural springs and *aflaj*) and lentic resources (man-made dams) [38]. Groundwater is accessed via wells established by a government agency which supplies the water to the communities through networks or water trucks. These wells are the main local source of water for drinking and domestic purposes (municipal, commercial) in the area. *Aflaj* are surface and/or underground channels fed by groundwater or a spring, or streams, built to provide water to the farming communities. The *aflaj* water is managed and distributed to farming areas by local people with no involvement of government in their organizational structure. Dams are artificial structures which are constructed by the government to harvest rainy water. *Aflaj* and dam water are mainly used for agriculture and livestock.

Al Jabal Al Akhdar has limited and highly variable water supplies: the most significant parameters influencing freshwater availability and causing environmental stress are the amount and frequency of rainfall. According to the climate change projection for the country [39], the variability of rainfall is expected to further increase, adding more uncertainty and complication to the planning and management of water resources. Furthermore, the population of the area has grown rapidly, from less than 2000 in 1970 to over 7000 in 2010 [40]. Socioeconomic development and related land-use changes due to the expansion of infrastructure and services, construction and commercial activities and urbanization have direct impacts in terms of increasing demand for water resources [37]. The establishment of settlements has been influenced strongly by the availability of water for drinking and domestic uses; all people have access to safe and good quality drinking water [37]. Together, the anthropogenic activities and climate change have affected the availability of water resources, and if these trends continue, the area's ecosystems and residence's households will be further affected. Vulnerability assessment of the environmental and climate change impacts on water resources is therefore essential to inform sustainable water resources management in the area.

3. Methodology

The methodological guidelines for "Vulnerability Assessment of Freshwater Resources to Environmental Change", developed by United Nations Environment Programme (UNEP) and Peking University [10] were used to assess the vulnerability of water resources of Al Jabal Al Akhdar to environmental change and climate impacts. According to the guidelines, the vulnerability of water resources can be assessed from two perspectives: the main threats to water resources and their development and utilization dynamics; and the region's challenges in coping with these threats. The threats can be assessed in terms of resource stresses (RS), development pressure (DP), ecological health (EH) and management capacity (MC). Thus, the vulnerability index (VI) of the water resources can be expressed as: $VI = f(RS, DP, EH, MC)$ [10].

Each component of VI has several parameters: $RS = f$ [water stress (RS_s) and water variation (RS_v)]; $DP = f$ [water exploitation (DP_s) and safe drinking water inaccessibility (DP_d)]; $EH = f$ [water pollution (EH_p) and ecosystem deterioration (EH_e)]; $MC = f$ [water use inefficiency (MC_e), improved sanitation inaccessibility (MC_s), and conflict management capacity (MC_g)]. In accordance with the vulnerability assessment guidelines, a number of governing equations were applied to estimate these parameters and VI (Table 1).

RS determines the water resources availability to meet the pressure of water demands for the growing population taking into consideration the rainfall variability. Therefore, it is influenced by the renewable water resources stress (RS_s) and water variation parameter resulting from long-term rainfall variability (RS_v). RS_s is expressed as per capita water resources and usually compared to the internationally agreed water poverty index of per capita water resources (1700 m^3/person/year) [10]. As Oman is part of West Asia, characterized by scarce water resources, the more appropriate and realistic value of 1000 m^3/person/year [9] was used. RS_v was estimated by the coefficient of variation (CV) of the rainfall record from 1979 to 2012, obtained from Saiq Meteorology Station. The CV was estimated by the ratio of the standard deviation of the rainfall record to the average rainfall (Table 1).

DP was estimated in terms of the overexploitation of water resources (DP_s) and the provision and accessibility of safe drinking water supply (DP_d). DP_s was estimated by the ratio of the total water demands (domestic, commercial, agriculture) to the total renewable water resources (Table 1). DP_d is defined here as the provision of adequate drinking water supplies to meet the basic needs for the society, in regard to how the water development facilities address the population needs [9]. The lack of safe water accessibility was estimated by the ratio of the percentage of population lacking accessibility to the size of the population (Table 1).

Table 1. Equations used for calculation of all categories and parameters of vulnerability index of water resources in the study area.

Category	Parameter	Equation	Description
Resource Stress (RS)	RS_s	$RS_s = (1000 - R)/1000$	R: Total renewable water resources per capita (m^3/person/year)
	RS_v	$RS_v = CV/0.3$ $CV = S/\mu$	CV: Coefficient of variation μ: Mean rainfall (mm) S: Standard deviation
Development Pressures (DP)	DP_s	$DP_s = WRs/WR$	WRs: Total water demands WR: Total renewable water resources
	DP_d	$DP_d = P_d/P$	P_d: Population without access to improved drinking water sources P: Total population of the area
Ecological Health (EH)	EH_p	$EH_p = (WW/WR)/0.1$	WW: Total untreated wastewater WR: Total renewable water resources
	EH_e	$EH_e = A_d/A$	A_d: Land area without vegetation coverage A: Total area of the country
Management Capacity (MC)	MCe	$MC_e = (WE_{wm} - WE)/WE_{wm}$	WE: GDP value produced from 1 m^3 of water WE_{wm}: Mean WE of West Asia countries
	MC_s	$MC_s = P_d/P$	P_d: Population without access to improved sanitation P: Total population of the area
	MC_g	MC_g = parameter matrix	Matrix scoring criteria (Table 2)
$$VI = \sum_{i=1}^{n}\left[\left(\sum_{j=1}^{m_i} x_{ij}*w_{ij}\right)*w_i\right]$$			n: number of parameter category m_i: number of parameters in ith category X_{ij}: value of jth parameter in ith category W_{ij}: Weight given to jth parameter in ith category W_i: Weight given to ith category

EH was measured in terms of the water quality/water pollution parameter (EH_p) and the ecosystem deterioration parameter (EH_e). EH_p was estimated by the ratio of the total untreated wastewater discharge in water receiving systems to the total available renewable water resources (Table 1). The amount of untreated wastewater is estimated as the difference between the generated wastewater collected by the system and amount of wastewater that received treatment. EH_e is defined in this study as the ratio of land area without vegetation coverage (*i.e.*, total land area except that covered with pastures and cultivated areas) to the total land area of Oman (309,500 km^2) (Table 1).

MC assesses the vulnerability of water resources by evaluating the current management capacity to cope with three critical issues: efficiency of water resources use; human health in relation to accessibility to adequate and safe sanitation services; and overall conflict management capacity. Thus, MC was measured with Water use inefficiency parameter (MC_e), Improved Sanitation inaccessibility parameter (MC_s), and Conflict Management Capacity Parameter (MC_g). MC_e was estimated in terms of the financial contribution to gross domestic product (GDP) of one cubic meter of water in any of the water consuming sectors compared to the world average for a selection of countries [10]. Since the agriculture sector is the major consumer of water in Oman, including the study area, it was used to indicate the financial return from the water use. Therefore, MC_e was calculated using US \$40 as the mean GDP value produced from 1 m^3 of water for the countries of West Asia [9] (Table 1). MC_s was used as a typical value to measure the capacity of the management system to deal with livelihood improvement in reducing pollution levels. Improved sanitation was defined here as facilities that hygienically separate human excreta from human, animal and insect contact, including sewers, septic tanks, flush toilets, latrines and simple pits [10]. MC_s was estimated as the ratio of proportion of the population without accessibility to improved sanitation facilities to the total population of the area (Table 1). MC_g demonstrates the capacity of a water resources management system to deal with conflicts. A good management system can be assessed by its effectiveness in institutional arrangements, policy formulation, communication mechanisms, and implementation efficiency [10]. The parameter was defined here as the capacity of the area to manage competition over water utilization among different consuming sectors. MC_g was determined based on water assessment survey [37] and expert consultation [41] using conflict management capacity scoring criteria ranging from 0.0 to 0.25 (Table 2), taking into consideration the interrelation of all variables in this table. These aspects were assigned scoring criteria ranging from 0 to 1 giving weights to each parameter.

Table 2. Conflict management capacity parameter assessment matrix (Source: [10]).

Category of Capacity	Description	Scoring Criteria		
		0.0	0.125	0.25
Institutional capacity	Trans-boundary institutional arrangement for coordinated water resources management	Solid institutional arrangements	Loose institutional arrangements	No existing institutions
Agreement capacity	Writing/signed policy/ agreement for water resources management	Concrete/detailed agreement	General agreement only	No agreement
Communication capacity	Routine communication mechanism for water resources management (annual conferences, *etc.*)	Communications at policy and operational levels	Communications only at policy level or operational level	No communication mechanism
Implementation capacity	Water resources management cooperation actions	Effective implementation of basin-wide river projects/programs	With joint project/ program, but poor management	No joint project program

Because the process of determining relative weights can be biased, making it difficult to compare the final results, equal weights were assigned among the parameters in the same category, and also among different categories. According to the guidelines [10], the weight of 0.25 was assigned across all categories (RS, DP, EH, and MC). For parameters RS_s, RS_v, DP_s, DP_d, EH_p and EH_e, the weight of 0.5 was applied, and for parameters MC_e, MC_s, and MC_g, the weight of 0.33 was assigned. The total weights given to all parameters in each category should be equal to 1, and the total weights given to all categories should be also equal to 1 [10].

The vulnerability index (VI) was finally estimated based on the four categories using the equation in Table 1. VI provides an estimated value ranging from zero (non-vulnerable) to one (most vulnerable) to determine the severity of the stress being experienced by the water resources of the study area. A high VI value shows high resource stresses, development pressures and ecological health, and low management capacities.

4. Results and Discussion

4.1. Resource Stresses

4.1.1. Water Stress Parameter

The calculation of water stress for Oman, including the study area, shows a critical water stress (RS_s = 0.58) (Table 3) based on the estimated total renewable water resources per capita of 422.5 m^3/person/year [42]. The increase in population and rapid socioeconomic development in Al Jabal Al Akhdar exert pressures on water resources: domestic water consumption increased from 150,000 m^3 in 2001 to 580,000 m^3 in 2012; an annual increase of 35% per year [37]. Much of this increase may be due to the burgeoning tourist industry. For 1985, 1995, and 2005, the calculated RS_s for Oman were 0.0, 0.30, and 0.36 based on the estimated per capita renewable water resources of 1029.35, 697.76 and 635.84 m^3/person/year, respectively [9].

Table 3. Calculated Vulnerability Index with various categories and parameters for the water resources of the study area.

Category	Resource Stress		Development Pressure		Ecological Health		Management Capacity		
Parameter	RS_s	RS_v	DP_s	DP_d	EH_p	EH_e	MC_e	MC_s	MC_g
Calculated	0.580	0.330	0.210	0.000	0.140	0.940	1.000	0.000	0.950
Weight in Category	0.50	0.50	0.50	0.50	0.50	0.50	0.33	0.33	0.33
Weighted	0.290	0.165	0.105	0.000	0.070	0.470	0.330	0.000	0.314
Component Total	0.4550		0.1050		0.5400		0.6435		
Weight for Category	0.25		0.25		0.25		0.25		
Weighted	**0.1138**		**0.0263**		**0.1350**		**0.1609**		
Overall Score	**0.436 (High)**								

Notes: Water Stress (RS_s); Water Variation (RS_v); Water Exploitation (DP_s); Safe Drinking Water Inaccessibility (DP_d); Water Pollution (EH_p); Ecosystem Deterioration (EH_e); Water Use Inefficiency (MC_e); Improved Sanitation Inaccessibility (MC_s); Conflict Management Capacity (MC_g).

4.1.2. Water Variation Parameter

Rainfall amount and availability are the dominant factors in the supply of water resources in the study area. Analysis of rainfall data records from 1979 to 2012 resulted in a water variation parameter (RS_v) of 0.33, based on the estimated CV of 0.10, indicating low rainfall variability. The methodology guidelines [9] designate a set of rainfall variation values for the coefficient of variation as CV = 0.3 or as a CV > 0.3. When CV is > 0.3, RS_v is assigned a highest value of 1, indicating large rainfall variation in time and space; a CV less than 0.3 reflects low variability. However, the study area experienced increasing temperatures over the same period (Figure 2). Minimum, mean and maximum temperatures increased at rates of 0.79, 0.27 and 0.15 °C per decade, respectively. Analysis of rainfall data showed a reduction in water availability, with a general decrease in total rainfall from 1979 to 2012 (Figure 2). Over this period, the average rainfall was 296.7 mm; the highest total was in 1997 (901 mm) and annual rainfall decreased subsequently to 202.8 mm in 2012, with an overall decrease in total rainfall at a rate of −9.42 mm per decade; indicating that the area is vulnerable to climate change as it is an arid mountain region. Projection of future climate in Oman using the IPCC A1B scenario shows an increase in temperature and a decrease in rainfall over the coming decades [39].

Figure 2. Trends in mean air temperature (Tmean) and annual rainfall in Saiq Meteorology Station (World Meteorological Organization (WMO) Index: 41254, Universal Transverse Mercator (UTM) coordinates Latitude: 23°04'28.33" N, Longitude: 57°38'46.63" E, Elevation: 1986 m) from 1979 to 2012 (Data source: [33]).

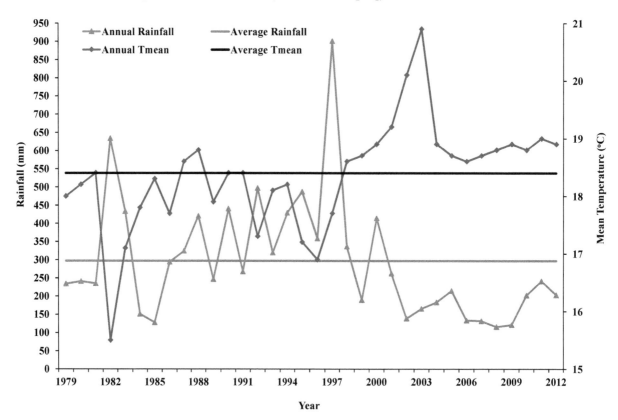

4.2. Water Development Pressures

4.2.1. Water Exploitation Parameter

The assessment of water development pressures indicated that the study area suffers from critical conditions in the development of water resources as determined by the water exploitation parameter (DP_s = 0.21) based on total water demands of 14 million m^3/year and the available total water resources of 66 million m^3/year [35] (Table 3), resulting in water shortages for domestic and agricultural purposes. There have been increases in the total population and socioeconomic development as well as increases in construction and commercial activities including hotels, and therefore water consumption by different sectors, causing an imbalance between supply and demand in the absence of the implementation of any conservation and management practices.

4.2.2. Safe Drinking Water Inaccessibility Parameter

The calculated safe drinking water inaccessibility parameter (DP_d) was zero since the fundamental needs of the population for water to live are met. There is sufficient infrastructure for providing drinking water throughout the study area; all people have access to safe drinking water. The government supplies drinking water to all households via groundwater wells, and a piped desalinated water project is in progress, to increase the availability of drinking water in the area.

4.3. Ecological Health

4.3.1. Water Pollution Parameter

The estimated water pollution parameter value was ($EH_p = 0.14$) (Table 3) based on the total untreated wastewater of 945,250 m^3/year [43] and the total available water resources of 66 million m^3/year, given that the urban water usage is 1.1 million m^3/year [35]. The analysis indicates low water pollution risks, which may be attributed to investments in wastewater treatment facilities: the government has established three wastewater treatment plants in the area with tertiary treatment levels and some sewerage systems, and all modern houses and other establishments have septic tanks. However, more investments are needed to increase the proportion of sewer networks connected with the treatment plants. Moreover, some septic tanks in old houses have unlined foundations [43], and need to be reconstructed to avoid pollution to groundwater aquifers.

4.3.2. Ecosystem Deterioration Parameter

Ecosystem deterioration due to the absence of adequate vegetation cover and modified natural landscape is a critical parameter in Oman including the study area, causing severe problems in supporting the functioning of ecosystems. EH_e was calculated as 0.94, based on the evaluation report of the land degradation and desertification in Arab Region [44] including Oman, as there is no available data on ecosystem deterioration for Al Jabal Al Akhdar. There are some indications of ecosystem deterioration in the study area due to decreased rainfall over the last three decades and therefore a decline of groundwater levels and the drying up of most *aflaj* [37]. The population growth, associated with anthropogenic activities and socioeconomic development, and overgrazing, as well as water overconsumption and expansion of land uses through sustained urbanization, have contributed directly or indirectly to the vulnerability of the water resources. The world map of the status of human-induced soil degradation [45] shows that the primary factor contributing to soil degradation in the Al-Hajar Mountains is loss of topsoil through water erosion, with 25%–50% of the area affected by a moderate degree of degradation. According to the study of desertification in the Arab Region by ACSAD (1997) as reported by [46], 89% of Oman was considered as desertified and 7.67% as vulnerable to desertification.

4.4. Management Capacity

4.4.1. Water Use Inefficiency Parameter

Based on the 2013 GDP of Oman (US\$80.6 billion) [47] and the total water withdrawal of 1321 million m^3/year [42], the calculated water use inefficiency (MC$_e$) was zero. This is in agreement with [9] which concluded that Oman showed the greatest efficiency gains (decreasing inefficiency) in the West Asia region between 1985 and 2005 (decrease of 25.37%). This was attributed to the uptake of more modern and efficient irrigation infrastructure systems.

However, this parameter was not calculated for the study area since it is based on the country scale and cannot be estimated at a regional scale. In Al Jabal Al Akhdar, farmers still use a traditional method of irrigation by flooding, with no application of modern irrigation technology or investments in improving irrigation infrastructure systems. Based on water assessment survey [37] and personal communication [41] with the author of the UNEP report [9] on this situation, MC$_e$ for the study area was estimated as 1, representing high water use inefficiency. This indicates unsustainable water resources management practices in the absence of a comprehensive water sector plan and strategy, leading to reduced water availability and increased vulnerability.

4.4.2. Improved Sanitation Inaccessibility Parameter

The entire population of the study area has access to sanitation facilities, such as sewer systems, septic tanks and wastewater treatment plants (MC$_s$ = 0) (Table 3), indicating adequate management regarding livelihood improvement through government investment in sanitation infrastructure. The availability of this infrastructure reduces pollution levels and preserves water resources, complemented by the implementation of policies and measures which may reduce the vulnerability of water resources to environmental and climate changes. However, more investments to expand the sewerage systems, connecting all households and other establishments to the wastewater treatment plants, are needed.

4.4.3. Conflict Management Capacity Parameter

The study area has no competition over water utilization with the neighboring regions. However, there is competition over water utilization between different sectors (agriculture and domestic). Agriculture is the dominant water consumer, with no application of conservation mechanisms and proper management capacity. There is also an increase in the domestic water consumption from groundwater wells, due to an increase in population and number of hotels and commercial activities, and there is no clear strategy for the development of the area [37]. Therefore, the assessment of MC$_g$ showed a high vulnerability situation in regard to conflict management capacity (MC$_g$ = 0.95) since this parameter takes into consideration the interrelation of different categories including institutional, agreement, communication and implementation capacity.

4.5. Vulnerability Index

Based on the available data, the calculated VI is 0.436, in the range of 0.4–0.7 which is classified as high based on the reference sheet for the interpretation of VI [10], indicating that the water resources of Al Jabal Al Akhdar are highly vulnerable and experiencing high stresses. Ecosystem deterioration is the dominant parameter, contributing 27% (Figure 3a). The area has also been experiencing a high degree of water use inefficiency, conflict management capacity and water stress representing 19%, 18% and 17%, respectively (Figure 3a), influencing the overall vulnerability on water resources. Comparison of the share of the different category groups to the final VI showed that the management capacity contributes most to the water resources vulnerability and is the dominant category (37%), followed by ecological health with 31% and water resources stress with 26% (Figure 3b).

Figure 3. (**a**) Percentage of the weighted parameters for Vulnerability Index; (**b**) Share of the percentage of the weighted categories to the final Vulnerability Index for the study area.

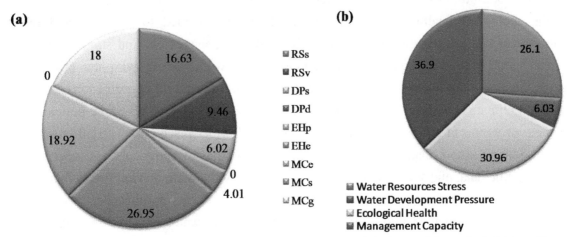

Notes: Water Stress (RS_s); Water Variation (RS_v); Water Exploitation (DP_s); Safe Drinking Water Inaccessibility (DP_d); Water Pollution (EH_p); Ecosystem Deterioration (EH_e); Water Use Inefficiency (MC_e); Improved Sanitation Inaccessibility (MC_s); Conflict Management Capacity (MC_g).

5. Conclusions and Recommendations

This is the first comprehensive vulnerability assessment of water resources in Al Jabal Al Akhdar, or Oman. The results have served to highlight which aspects of water management (resources stress, development pressure, ecological health, and management capacity) contribute most to the vulnerability of water resources and to understand the various risks and thus to suggest potential areas to best focus management efforts. The vulnerability assessment indicated high VI (0.436). Ecosystem deterioration is the dominant parameter contributing 27% to the vulnerability index. The water resources of the area have also been experiencing a high degree of water use inefficiency, conflict management capacity and water stress, influencing the overall vulnerability index by 19%, 18% and 17%, respectively. Management capacity is the dominant category, representing 37% of the category groups, driving the vulnerability of the water resources, which are also highly influenced by the ecological health (31%) and water resources stress (26%). These could be used as indicators for the vulnerability of water resources to environmental and climate changes in the study area. Nevertheless, it must be recognized that due to the lack of availability of local data, some of the inputs to the assessment are at national scale.

There is a clear need for policies and technical solutions to mitigate the pressures (water over consumption and inefficient use, ecosystem deterioration, climate change) which make the water resources more vulnerable. A longer term strategic development plan should be made, with a focus on management capacity to deal with the main threats of conflicts between water consuming sectors, as well as implementation of effective management practices in line with the integrated water resources management approach. Additional effort is needed to improve irrigation water use efficiency, conservation technologies, rainwater harvesting, and reuse of treated wastewater and grey water to relieve some of the agricultural pressures on water resources. There is also an urgent need for mitigation and adaptation to climate change impacts since the region is expected to face further increases in temperatures and decreases in rainfall over the coming decades.

The major contribution of ecosystem deterioration to the overall index suggests that, in order to sustain the ecological health of the area, more efforts are needed to conserve and rehabilitate vegetation cover and implement best practices for land use management and strategic development. More investments are also required to expand sewer networks along with the effective use of wastewater treatment facilities to protect freshwater from pollution. Full coordination, integration and awareness on climate change adaptation should be strongly connected to planning, policies and water management programs at all levels and across all sectors. Further research is needed to provide local, rather than national, input data particularly on the deterioration of ecosystem and vegetation cover, long-term climatic data, and socioeconomic trends, to identify the main driving forces that increase the vulnerability of the water resources, in order to define the optimal approaches for climate change adaptation, to be implemented into operational and sustainable water resources management for the green mountain.

Acknowledgments

The authors would like to thank the Centre for Mountain Studies, Perth College, The University of the Highlands and Islands for covering the cost to publish this article in an open access journal. The authors would also like to thank the Directorate General of Meteorology and Air Navigation, Public Authority of Civil Aviation, Muscat, Oman, for providing rainfall and temperature data for the study area. Furthermore, the authors greatly appreciate the careful revisions of the reviewers.

Author Contributions

Mohammed Al Kalbani collected and analyzed the data, and applied the methodology in collaboration with Martin Price and Asma Abahussain. He also performed all calculations and checked the results in collaboration with Mushtaque Ahmed and Timothy O'Higgins. In addition, he evaluated the data and interpretation the results with Asma Abahussain, and enhanced the writing and editing of the paper with Martin Price.

References

1.   Millennium Ecosystem Assessment (MEA). *Ecosystem and Human Wellbeing: Current State and Trends*; Island Press: Washington, DC, USA, 2005.

2. Intergovernmental Panel on Climate Change (IPCC). Summary for Policymakers. In *Climate Change 2007: Impacts, Adaptation and Vulnerability*; Contribution of Working Group II to the Fourth Assessment Report of the Intergovernmental Panel on Climate Change; Parry, M.L., Canziani, O.F., Palutikof, J.P., van der Linden, P.J., Hanson, C.E., Eds.; Cambridge University Press: Cambridge, UK, 2007; pp. 7–22.

3. Intergovernmental Panel on Climate Change (IPCC). Summary for Policymakers. In *Climate Change 2007: The Physical Science Basis*; Contribution of Working Group I to the Fourth Assessment Report of the Intergovernmental Panel on Climate Change; Solomon, S., Qin, D., Manning, M., Chen, Z., Marquis, M., Averyt, K.B., Tignor, M., Miller, H.L., Eds.; Cambridge University Press: Cambridge, UK, 2007; pp. 1–18.

4. Gosling, S.N.; Warren, R.; Arnell, N.W.; Good, P.; Caesar, J.; Bernie, D.; Lowe, J.A.; van der Linden, P.; O'Hanley, J.R.; Smith, S.M. A review of recent developments in climate change science. Part II: The global-scale impacts of climate change. *Prog. Phys. Geogr.* **2011**, *35*, 443–464.

5. Intergovernmental Panel on Climate Change (IPCC). Climate Change 2014. In *Mitigation of Climate Change*; Contribution of Working Group III to the Fifth Assessment Report of the Intergovernmental Panel on Climate Change; Edenhofer, O., Pichs-Madruga, R., Sokona, Y., Farahani, E., Kadner, S., Seyboth, K., Adler, A., Baum, I., Brunner, S., Eickemeier, P., *et al.*, Eds.; Cambridge University Press: Cambridge, UK and New York, NY, USA, 2014.

6. Kundzewicz, Z.W.; Mata, L.J.; Arnell, N.W.; Döll, P.; Kabat, P.; Jiménez, B.; Miller, K.A.; Oki, T.; Sen, Z.; Shiklomanov, I.A. Freshwater resources and their management. In *Climate Change: Impacts, Adaptation and Vulnerability*; Contribution of Working Group II to the Fourth Assessment Report of the Intergovernmental Panel on Climate Change; Parry, M.L., Canziani, O.F., Palutikof, J.P., van der Linden, P.J., Hanson, C.E., Eds.; Cambridge University Press: Cambridge, UK, 2007; pp. 173–210.

7. Rockstrom, J.; Falkenmark, M.; Karlberg, L.; Hoff, H.; Rost, S.; Gerten, D. Future water availability for global food production: The potential of green water for increasing resilience to global change. *Water Resour. Res.* **2009**, *45*, 1–16.

8. Gain, A.K.; Giupponi, C.; Renaud, F.G. Climate Change Adaptation and Vulnerability Assessment of Water Resources Systems in Developing Countries: A Generalized Framework and a Feasibility Study in Bangladesh. *Water* **2008**, *4*, 345–366.

9. United Nations Environment Programme (UNEP). *Assessment of Freshwater Resources Vulnerability to Climate Change: Implication on Shared Water Resources in West Asia Region*; UNEP: Nairobi, Kenya, 2012; pp. 1–164.

10. Huang, Y.; Cai, M. *Methodologies Guidelines: Vulnerability Assessment of Freshwater Resources to Environmental Change*; United Nations Environment Programme (UNEP) and Peking University, China; UNEP, Regional Office for Asia and the Pacific: Bangkok, Thailand, 2009; pp. 1–28.

11. Ribot, J. The causal structure of vulnerability: Its application to climate impact analysis. *GeoJournal* **1995**, *35*, 119–122.

12. Downing, T.E.; Patwardhan, A. *Vulnerability Assessment for Climate Adaptation*; APF Technical Paper 3, Final Draft; United Nations Development Programme: New York, NY, USA, 2003.

13. Bankoff, G.; Frerks, G.; Hilhorst, D. *Mapping Vulnerability: Disasters, Development and People*; Earthscan: London, UK, 2004.

14. O'Brien, K.; Eriksen, S.; Schjolden, A.; Nygaard, L. *What's in a Word? Conflicting Interpretations of Vulnerability in Climate Change Research*; CICERO Working Paper 2004:04; Center for International Climate and Environmental Research: Oslo, Norway, 2004.

15. Brooks, N.; Adger, W.N.; Kelly, P.M. The determinants of vulnerability and adaptive capacity at the national level and the implications for adaptation. *Glob. Environ. Chang.* **2005**, *15*, 151–163.

16. Adger, W.N. Social capital, collective action, and adaptation to climate change. *Econ. Geogr.* **2003**, *79*, 387–404.

17. Adger, W.N.; Huq, S.; Brown, K.; Conway, D.; Hulme, M. Adaptation to climate change in the developing world. *Prog. Dev. Stud.* **2004**, *3*, 179–195.

18. Eakin, H.; Luers, A. Assessing the vulnerability of social-environmental systems. *Annu. Rev. Environ. Resour.* **2006**, *31*, 365–394.

19. Brooks, N. *Vulnerability, Risk and Adaptation: A Conceptual Framework*; Working Paper 38; Tyndall Center for Climate Change Research: Norwich, UK, 2003.

20. Kelly, P.M.; Adger, W.N. Theory and practice in assessing vulnerability to climate change and facilitating adaptation. *Clim. Chang.* **2000**, *47*, 325–352.

21. Wisner, B.; Blaikie, P.; Cannon, T.; Davis, I. *At Risk: Natural Hazards, People's Vulnerability and Disasters*; Routledge: London, UK, 2004.

22. Füssel, H.M.; Klein, R.J.T. Climate change vulnerability assessments: An evolution of conceptual thinking. *Clim. Chang.* **2006**, *75*, 301–329.

23. Eakin, H. The social vulnerability of irrigated vegetable farming households in central Puebla. *J. Environ. Dev.* **2003**, *12*, 414–429.

24. Adger, W.N. Vulnerability. *Glob. Environ. Chang.* **2006**, *16*, 268–281.

25. O'Brien, K.; Eriksen, S.; Nygaard, L.; Schjolden, A. Why different interpretations of vulnerability matter in climate change discourses. *Clim. Policy* **2007**, *7*, 73–88.

26. Schneider, S.H.; Semenov, S.; Patwardhan, A.; Burton, I.; Magadza, C.H.D.; Oppenheimer, M.; Pittock, A.B.; Rahman, A.; Smith, J.B.; Suarez, A.; Yamin, F. Assessing Key Vulnerabilities and the Risk from Climate Change. In *Climate Change 2007: Impacts, Adaptation and Vulnerability*; Contribution of Working Group II to the Fourth Assessment Report of the Intergovernmental Panel on Climate Change; Parry, M.L., Canziani, O.F., Palutikof, J.P., van der Linden, P.J., Hanson, C.E., Eds.; Cambridge University Press: Cambridge, UK, 2007; pp. 779–810.

27. Adger, W.N.; Agrawala, S.; Mirza, M.M.Q.; Conde, C.; O'Brien, K.; Pulhin, J.; Pulwarty, R.; Smit, B.; Takahashi, K. Assessment of Adaptation Practices, Options, Constraints and Capacity. In *Climate Change 2007: Impacts, Adaptation and Vulnerability*; Contribution of Working Group II to the Fourth Assessment Report of the Intergovernmental Panel on Climate Change, Parry, M.L., Canziani, O.F., Palutikof, J.P., van der Linden, P.J., Hanson, C.E., Eds.; Cambridge University Press: Cambridge, UK, 2007; pp. 717–743.

28. McCarthy, J.J.; Canziani, O.F.; Leary, N.A.; Dokken, D.J.; White, K.S. *Climate Change 2001: Impacts, Adaptation, and Vulnerability*; Cambridge University Press: Cambridge, UK, 2001.

29. Lavell, A.; Oppenheimer, M.; Diop, C.; Hess, J.; Lempert, R.; Li, J.; Muir-Wood, R.; Myeong, S. Climate Change: New Dimensions in Disaster Risk, Exposure, Vulnerability, and Resilience. In

Managing the Risks of Extreme Events and Disasters to Advance Climate Change Adaptation; A Special Report of Working Groups I and II of the Intergovernmental Panel on Climate Change (IPCC); Field, C.B., Barros, V., Stocker, T.F., Qin, D., Dokken, D.G., Ebi, K.L., Mastrandrea, M.D., Mach, K.J., Plattner, G.K., Allen, S.K., *et al.*, Eds.; Cambridge University Press: Cambridge, UK, and New York, NY, USA, 2012; pp. 25–64.

30. Turner, B.L.; Kasperson, R.E.; Matson, P.A.; McCarthy, J.J.; Corell, R.W.; Christensen, L.; Eckley, N.; Kasperson, J.X.; Luers, A.; Martello, M.L.; *et al.* A framework for vulnerability analysis in sustainability science. *Proc. Natl. Acad. Sci. USA* **2003**, *100*, 8074–8079.

31. Ionescu, C.; Klein, R.J.T.; Hinkel, J.; Kavi Kumar, R.S.; Klein, R. Towards a Formal Framework of Vulnerability to Climate Change. Available online: www.usf.uni-osnabrueck.de/projects/newater/downloads/newater_wp02.pdf (accessed on 9 September 2014).

32. Luedeling, E. Sustainability of Mountain Oases in Oman: Effects of Agro-Environmental Changes on Traditional Cropping Systems. Master's Thesis, The University of Kassel, Kassel, Germany, 2007.

33. *Director General of Meteorology and Air Navigation (DGMAN)*; Public Authority of Civil Aviation: Muscat, Oman, 2014.

34. Al-Riyami, Y. *Agriculture Development in Al Jabal Al Akhdar*; Working paper presented in the "Symposium of Economic Development in Al Jabal Al Akhdar"; Oman Chamber of Commerce and Industry: Nizwa Branch, Oman, 2006. (In Arabic)

35. MacDonald, M. *Water Balance Computation for the Sultanate of Oman*; Ministry of Regional Municipality and Water Resources: Muscat, Oman, 2013.

36. Patzelt, A. Syntaxonomy, Phytogeography and Conservation Status of the Montane Flora and Vegetation of Northern Oman - A Centre of Regional Biodiversity. In Proceedings of the International Conference on Mountains of the World: Ecology, Conservation and Sustainable Development, Muscat, Sultanate of Oman, 10–14 February 2008; Victor, R., Robinson, M.D., Eds.; Sultan Qaboos University: Muscat, Sultanate of Oman, 2009.

37. Al Kalbani, M.S. Integrated Environmental Assessment and Management of Water Resources in Al Jabal Al Akhdar Using the DPSIR Framework, Policy Analysis and Future Scenarios for Sustainable Development. Ph.D. Thesis, University of Aberdeen, Aberdeen, UK, in progress.

38. Victor, R.; Ahmed, M.; Al Haddabi, M.; Jashoul, M. Water Quality Assessments and Some Aspects of Water Use Efficiency in Al Jabal Al Akhdar. In Proceedings of the International Conference on Mountains of the World: Ecology, Conservation and Sustainable Development, Muscat, Sultanate of Oman, 10–14 February 2008; Victor, R., Robinson, M.D., Eds.; Sultan Qaboos University: Muscat, Sultanate of Oman, 2009.

39. Al-Charaabi, Y.; Al-Yahyai, S. Projection of Future Changes in Rainfall and Temperature Patterns in Oman. *J. Sci. Clim. Chang.* **2013**, *4*, 154–161.

40. National Centre for Statistics and Information (NCSI). *Census 2010: Final Results, General Census on Population, Housing & Establishments 2010*; NCSI: Muscat, Oman, 2012.

41. Abahussain, A.A. Department of Natural Resources and Environment, College of Graduate Studies, Arabian Gulf University, Kingdom of Bahrain, Personal communication, 2014.

42. Food and Agriculture Organization of the United Nations (FAO). AQUASTAT: FAO's Global Water Information System. Available online: http://www.fao.org/nr/water/aquastat/data/query/results.html (accessed on 14 September 2014).

43. Ministry of Regional Municipalities and Water Resources (MRMWR). *Data on Wastewater Treatment Plants and Sewer Networks*; Unpublished Data, 2014.

44. Arab Center for the Study of Arid Zones and Dry Lands (ACSAD); Council of Arab Ministers Responsible for the Environment (CAMRE); United Nations Environment Programme (UNEP). *State of Desertification in the Arab World (Updated Study)*; Arab Center for the Study of Arid Zones and Dry Lands: Damascus, Syria, 2004. (In Arabic).

45. Oldeman, L.R.; Hakkeling, R.T.A; Sombroek, W.G. *World Map of the Status of Human-induced Soil Degradation*; Global Assessment of Soil Degradation (GLASOD), International Soil Reference and Information Centre, United Nations Environment Programme: Nairobi, Kenya, 1991.

46. Abahussain, A.A.; Abdu, A.Sh.; Al-Zubari, W.K.; El-Deen, N.A; Abdul-Raheem, M. Desertification in the Arab Region: analysis of current status and trends. *J. Arid Environ.* **2002**, *51*, 521–545.

47. The World Bank. Data. Available online: http://data.worldbank.org/indicator/NY.GDP.MKTP.CD (accessed on 14 September 2014).

5

Runoff and Sediment Yield Variations in Response to Precipitation Changes

Tianhong Li [1,2,*] **and Yuan Gao** [1,2]

[1] College of Environmental Sciences and Engineering, Peking University, Beijing 100871, China;
E-Mail: gaoyuan@iee.pku.edu.cn
[2] Key Laboratory of Water and Sediment Sciences, Ministry of Education, Beijing 100871, China

* Author to whom correspondence should be addressed; E-Mail: lth@pku.edu.cn;

Academic Editors: Yingkui Li and Michael A. Urban

Abstract: The impacts of climate change on hydrological cycles and water resource distribution is particularly concerned with environmentally vulnerable areas, such as the Loess Plateau, where precipitation scarcity leads to or intensifies serious water related problems including water resource shortages, land degradation, and serious soil erosion. Based on a geographical information system (GIS), and using gauged hydrological data from 2001 to 2010, digital land-use and soil maps from 2005, a Soil and Water Assessment Tool (SWAT) model was applied to the Xichuan Watershed, a typical hilly-gullied area in the Loess Plateau, China. The relative error, coefficient of determination, and Nash-Sutcliffe coefficient were used to analyze the accuracy of runoffs and sediment yields simulated by the model. Runoff and sediment yield variations were analyzed under different precipitation scenarios. The increases in runoff and sediment with increased precipitation were greater than their decreases with reduced precipitation, and runoff was more sensitive to the variations of precipitation than was sediment yield. The coefficients of variation (Cvs) of the runoff and sediment yield increased with increasing precipitation, and the Cv of the sediment yield was more sensitive to small rainfall. The annual runoff and sediment yield fluctuated greatly, and their variation ranges and Cvs were large when precipitation increased by 20%. The results provide local decision makers with scientific references for water resource utilization and soil and water conservation.

Keywords: SWAT model; precipitation; runoff; sediment yield; simulation; the Xichuan River; the Loess Plateau

---

1. Introduction

Climate change as a result of both natural factors and human activities is altering the earth's hydrologic cycles to various degrees [1,2]. Climate change affects hydrology mainly through changes in precipitation, temperature, and evaporation [3,4], and it subsequently influences the temporal-spatial distributions of runoff and sediment, as well as the patterns of runoff and sediment transport [5]. The impacts of climate change on water resources and the hydrologic cycles have long been a focus of the international community [6,7]. Research on this issue began as early as the 1980s. In 1985, the World Meteorological Organization (WMO) published a summary report of their study of the impacts of climate change on hydrology and water resources and proposed several evaluation and test methods. In 1987, the WMO proposed analyzing the sensitivity of hydrology and water resources to climate change. This issue was also discussed in the 2007 international conference of the International Union of Geodesy and Geophysics (IUGG). The Intergovernmental Panel on Climate Change (IPCC) of the United Nations analyzed the impacts of climate change on hydrology and water resources from 1990 to 2007. In its technical report [8], the IPCC highlighted that the global and regional water resource problems caused by climate change are crucial issues. Changes in precipitation and temperature have significant effects on runoff and water availability, particularly in semiarid and arid regions [9].

China has always considered the impacts of climate change on water resources to be important and has actively carried out a series of scientific studies to support research on the impact of the changing environment (due to global changes and human activities) on water cycles [5,10]. For instance, the *National Planning Outline for Mid- and Long-term Scientific and Technological Development (2006–2020)* issued by the State Council of China in 2006 pointed out that research on the impacts of global climate change in China is a focus, with special emphasis on the impacts of climate change on hydrologic cycles and regional water resources, especially in arid regions with fragile ecological environments [11].

Currently, studies of the impacts of climate change on runoff and sediment mainly focus on two aspects. Some studies analyze the changes in the temporal-spatial distributions of runoff and sediment and the patterns of runoff and sediment transport that are caused by changes in climate factors, such as precipitation and temperature, whereas other studies analyze the trends of the changes in runoff and sediment under future climate change scenarios. The main method for quantitatively evaluating and studying the impacts of climate change on runoff and sediment is to use watershed hydrological models. The most commonly used models are statistical regression models, water balance models, and distributed physical models [9,12]. Of these models, the Soil and Water Assessment Tool (SWAT) [13], which was developed by the US Department of Agriculture in the 1990s, has been widely applied to watersheds around the world [14–20]. There are two types of predicted future climate change scenarios. First, changes in temperature, precipitation, and evaporation are hypothesized based on the trends and ranges of the meteorological changes in the study area, as well as specialized knowledge, experience, and the time-series statistical analysis method, which is easy to design and apply [21–23]; Second, different climate change scenarios can be simulated using models, such as the General Circulation Models [19,24].

Previous studies have shown that the precipitation in the Yellow River Basin has decreased significantly since the 1970s [10,25] although variation trends may differ in sub-basins. Precipitation is the main source of runoff and one of the driving factors of soil and water losses in the Loess Plateau [26,27], where water resources are scarce. Therefore, in the context of global climate change, studying the impacts of changes in precipitation on the runoff and sediment production in the Yellow River Basin is important for the sustainable utilization of water resources. Most previous studies of the impacts of changes in precipitation on water resources in the Yellow River Basin have focused on the entire basin [28,29] or the basin at relatively large scales [30–32]. Relatively few studies have focused on small watersheds. In addition, most studies have focused on the impact of precipitation on the runoff and have rarely investigated the impact of precipitation on the sediment yield. In fact, high sediment content is an important and unique characteristic of the rivers in the Loess Plateau, China. Sediment transport requires a considerable amount of water [33] and competes with other water uses. Thus, it is imperative to consider sediment when studying the water resource problems of these rivers.

This study used the Xichuan Watershed, a typical small basin in the hilly and gully area in the Loess Plateau, as a target area, and used ArcGIS, MATLAB, and SPAW [34,35] to process observed meteorological and hydrological data. Then, a localized SWAT model was constructed, calibrated, and validated. Using the SWAT model under different precipitation scenarios, the study quantitatively predicted the impacts of changes in precipitation on the runoff and sediment yield in this small watershed and analyzed the characteristics of the changes in runoff and sediment production with the aim of providing a scientific basis for the management and sustainable utilization of water resources in basins that are similar to the study area.

2. Methodology

Based on the spatial and attribute data, including meteorological data, hydrological data, soil map, land use map, and a digital elevation model (DEM), this study investigated the characteristics of the changes in the precipitation, runoff, and sediment yield in the study area. Spatial and attribute databases of the SWAT model were developed. After determining the parameters of the SWAT model and verifying the predicted results from the model, we quantitatively analyzed the impacts of changes in precipitation on the runoff and sediment production in the study area using precipitation change scenarios. Figure 1 shows the technical workflow of the study.

This study used AVSWAT, developed by integrating the SWAT into ArcView for the analysis. AVSWAT has powerful spatial analysis and processing functions and is convenient to use. The SWAT model consists of three sub-models: the hydrological process sub-model, the soil erosion sub-model and the pollution load sub-model. This study mainly uses the hydrological process and the soil erosion sub-models.

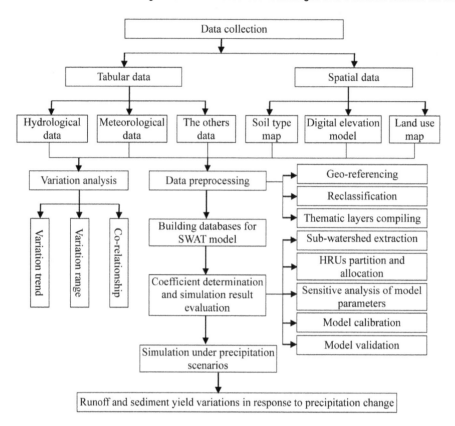

Figure 1. Technical workflow of this study.

2.1. Study Area

The Xichuan River is a tributary of the Yanhe River (a tributary of the Yellow River) with a total length of 61.5 km. The Xichuan Watershed is located west of the Yanhe River Basin between 108°50′ E and 109°20′ E and between 36°30′ N and 36°45′ N, covering an area of 801 km^2 [36]. The mean runoff of the watershed was 169.04 × 10^6 m^3 from 2001 to 2010. The river originates in Caofeng Village, Zhidan Town in Zhidan County and flows past Xihekou Village, Zhuanyaowan Town and Gaoqiao Village in Ansai County and Zaoyuan Village in the Baota District and eventually flows into the Yanhe River near Shifogou in the Baota District. The Zaoyuan Hydrological Station (ZHS) is located 13 km upstream from the mouth of the Xichuan River, and it controls 90% area (719 km^2) of the whole watershed (Figure 2).

The Xichuan Watershed has a continental monsoon climate where winters are cold and dry with little precipitation, whereas summers are warm with abundant precipitation. Precipitation is unevenly distributed and mainly concentrated in the summer and fall, accounting for 54.3% and 27.7% of the total annual precipitation. Floods in this watershed have relatively short durations, rising and falling suddenly with high sediment concentrations [37].

The soil types in this watershed include yellow loessial soil, red clay soil, alluvial soil, and dark loessial soil. The yellow loessial soil, developed from the parent loess, is the main soil type, covering more than 80% of the total basin area. The vegetation coverage of the watershed is very low, belonging to the forest steppe zone. Both natural and artificial vegetation types are present, mainly consisting of crops, evergreen coniferous forests, deciduous coniferous forests, deciduous broad-leaved forests, shrubs, and grasslands.

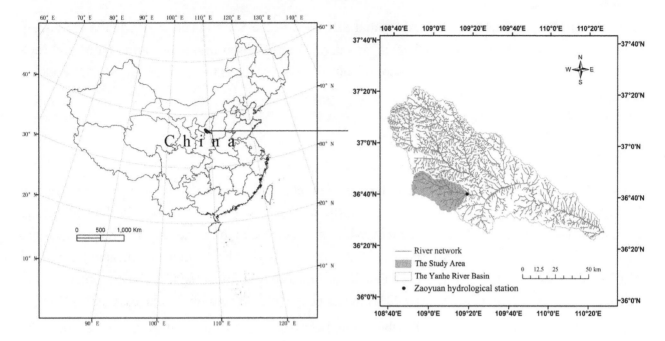

Figure 2. Location of the study area.

More than 80% of the basin area suffers soil erosion by water. The multi-year mean sediment discharge is 1330.2×10^4 t/a. Since the 1970s, a series of water and soil conservation and ecological construction projects have been implemented and have substantially improved the ecological environment in the watershed [38].

2.2. Data and Data Preprocessing

The main data used in this study included Digital Elevation Model (DEM) data, a land use map, a soil type map, precipitation data, temperature data, and the boundary of the Yanhe River basin. Table 1 lists the data descriptions and sources.

Table 1. List of data that were used in this study.

Data Type	Temporal/Spatial Resolution	Source
DEM data	Grid format, 30 m/grid	Data Application Environment Sharing Platform of the Chinese Academy of Sciences
Land use map	At the scale of 1:100,000, compiled at 2005	Data Application Environment Sharing Platform of the Chinese Academy of Sciences
Soil type map	At the scale of 1:1,000,000, compiled at 2005	Data Application Environment Sharing Platform of the Chinese Academy of Sciences
Meteorological data	Daily precipitation, daily maximum temperature and daily minimum temperature between 1990 and 2010	China Meteorological Data Sharing Service Website
Runoff and sediment yield	Monthly runoff and sediment yield between 2001 and 2010	The Zaoyuan Meteorological Station in the Yan'an City

The data required for the SWAT model included geospatial data, a non-spatial attribute database, meteorological data, and hydrological and sediment data for model calibration and verification. Spatial data processing, grid calculations, and interpolations were conducted using ArcGIS, and the statistical analysis was performed using Excel. All spatial data used in the SWAT were converted to the Albers equal-area conic projection. The whole area was divided into 31 sub-basins using the DEM and each sub-basin contained 5–16 hydrological response units (HRUs), which is the basic unit in SWAT model.

2.3. SWAT Model Development

2.3.1. Model Construction

The SWAT model requires the land use classification scheme developed by the US Geological Survey (USGS). It also requires the auxiliary land use attributes with parameters provided by the USGS. The land use map used in this study had to be reclassified to meet these requirements. After reclassification, the main land use types included farmland (39.30%), typical grassland (32.60%), meadow grassland (13.09%), deciduous coniferous forest and deciduous broad-leaved forest (8.90%), bush wood (5.05%), evergreen forest (0.86%), barren land (0.09%), water body (0.05%), and rural villages (0.05%) [35].

The soil data included the spatial distribution and physical and chemical attributes of the soils. The soil map was produced based on the 2005 soil survey of the study area. The physical attributes of the soils mainly included the thicknesses, silt contents, clay contents, bulk densities, organic carbon contents, effective water contents, saturated hydraulic conductivities, and the available field capacities. These attributes control the movement of the water and air in the soil and have an important role in the water cycles. This study established a users' soil parameter database based on these characteristics.

The soils were divided into four hydrologic groups. For the same precipitation and surface conditions, soils with similar runoff production capacities were classified into a single hydrologic group [39].

The wet density of a soil, the available effective water in a soil layer and the saturated hydraulic conductivity coefficient can be calculated using the SPAW model [34,35]. The SPAW model is a daily hydrologic budget model for agricultural fields. It also includes a routine for the daily water budgets of inundated ponds and wetlands that utilizes the field hydrology of the watershed.

The soil erosion factor (K) is often used to evaluate soil erodibility. K is calculated based on the organic carbon and particle compositions of the soil using the method proposed by Williams [40].

The observed meteorological input data mainly included the precipitation, daily maximum and minimum temperature data from 2001 to 2010. The SWAT model includes a built-in weather generator. If some data were not available, the weather generator simulated daily meteorological data based on multi-year monthly mean data that were provided in advance. The "pcpSTAT" and "dew02" procedures were used to calculate daily precipitation and temperature to obtain the related parameters and generate the weather data that were needed for simulations.

The measured runoff and sediment data were collected at ZHS (Figure 2) from 2001 to 2010. They were used in sensitivity analysis and parameters calibration. The measured daily precipitation data were used to simulate daily runoff using the Soil Conservation Service (SCS) curve method [39]. The potential evaporation was derived using the Penman-Monteith method [41]. The variable storage coefficient method [42] was used in the river channel routing simulation.

2.3.2. Sensitivity Analysis, Validation, and Testing of the SWAT Parameters

The parameter sensitivity analysis module was used to analyze the sensitivities of the parameters in the runoff and sediment simulations. This module uses the Latin hypercube one-factor-at-a-time (LH-OAT) method [43]. The objective of this analysis is to analyze and determine which input parameters have the most significant impacts on the output when their values are changed. Important parameters are selected to highlight their impacts on the simulation and to reduce the time that is needed for parameter adjustment. In this study, the simulated runoff and sediment yield were compared with actual gauged data at ZHS (Figure 2). The important factors affecting the precision of the simulation in the watershed were determined after analyzing the sensitivity of each parameter.

The runoff and sediment yield parameters were calibrated in sequence. Three indexes, including the relative error (Re) [44], the coefficient of determination (R^2) [45] and the Nash-Sutcliffe coefficient (Ens) [46] were chosen to statistically test the accuracy of the calibrated and validated runoff and sediment yield outputs. If $Re = 0$, the model prediction is the same as the observed data. If $Re > 0$, the model prediction is larger than the observation. If $Re < 0$, the model prediction is less than the observed value. R^2 was obtained from the linear regression in Microsoft Excel. The larger the R^2 value, the better simulation of the model. If the value of Ens is greater than 0.75, the simulation is excellent. If Ens is between 0.36 and 0.75, the simulation is satisfactory, and if Ens is less than 0.36, the simulation is unsatisfactory.

2.4. Precipitation Scenarios

Based on land use maps in 2005 and the climate conditions from 2001 to 2010, the calibrated and verified SWAT model was used to simulate the impacts of precipitation on the runoff and sediment yield by altering the input climate conditions (precipitation). Spatial variability in precipitation was not considered in the simulations because the study area is a small watershed with limited precipitation stations.

Precipitation scenarios were determined based on the variation characteristics of the precipitation in this area. During the study period, the annual precipitation did not show significant increasing or decreasing trend in this area. A previous study [47] also showed no significant increasing or decreasing trend in the Yanhe River basin. The mean annual precipitation in the study area from 2001 to 2010 was 514 mm. Thus, we considered four precipitation change scenarios:

(1) the annual mean precipitation increases by 20%, *i.e.*, 617 mm;
(2) the annual mean precipitation increases by 10%, *i.e.*, 565 mm;
(3) the annual mean precipitation decreases by 10%, *i.e.*, 462 mm; and
(4) the annual mean precipitation decreases by 20%, *i.e.*, 411 mm.

2.5. Variance Analysis

The coefficient of variation (Cv) was used to reflect the inter-annual changes in the precipitation and runoff. It is calculated using the following equation:

$$Cv = \frac{SD}{M} \qquad (1)$$

where *SD* is the standard deviation of a variable and *M* is the average value of the variable. The greater the value of *Cv* of precipitation or runoff, the greater the extent of the inter-annual change in the precipitation or runoff is, and the possibility of occurrences of floods or droughts increases. The smaller the value of *Cv* of the inter-annual precipitation or runoff is, the smaller the extent of the inter-annual change in the precipitation or runoff is, which is more beneficial to the utilization of water resources. *Cv* also reflects the characteristics of the inter-annual change in the sediment yield. The greater the value of *Cv* of the sediment yield indicates that the sediment yield changed greatly, and disasters such as soil erosion are more common. On the other hand, the smaller the value of *Cv* of sediment yield is, the smaller changing extent of the sediment yield is, which is more beneficial to water and soil conservation.

The changing rate (*CR*) is another parameter that expresses the changes in the runoff and sediment yield:

$$CR = \frac{X_i - X_0}{X_0} \times 100\% \tag{2}$$

where X_i represents the simulated annual mean runoff or sediment yield under the *i*th precipitation scheme, and X_0 represents the simulated annual mean runoff or sediment yield under the actual conditions.

3 Results

3.1. Characteristics of the Variations of Runoff and Sediment Yield

The Xichuan Watershed is in a continental monsoon climate region. The runoff mainly originates from precipitation and is thus significantly affected by precipitation. With the data measured at the ZHS between 2001 and 2010, the changes in the runoff and sediment yield in the watershed during this 10-year period were analyzed and compared with precipitation changes.

Figure 3 shows the distributions of the monthly mean precipitation and runoff between 2001 and 2010. The precipitation and runoff are both concentrated in the flood season (from June to September). The most intense monthly precipitation occurs in July, which accounts for 20.4% of the annual total, and the maximum monthly runoff occurs in August, accounting for 20.1% of the total annual runoff. The minimum monthly precipitation occurs in December (0.75% of the total annual precipitation), and the minimum monthly runoff occurs in January (3.38% of the total annual runoff). The seasonal precipitation and runoff distributions in the watershed are extremely uneven. Precipitation and runoff are mainly concentrated in the summer and fall. Summer precipitation accounts for 54.3% of the total annual precipitation, and summer runoff accounts for 45.7% of the total annual runoff. Fall precipitation accounts for 27.7% of the total annual precipitation, and fall runoff accounts for 23.3% of the total annual runoff. Figure 3 also shows that, during the spring and winter, the changes in the precipitation and runoff are gentle, and the precipitation and runoff are relatively stable. The runoff is slightly affected by precipitation during this period. Generally speaking, the runoff in the Xichuan Watershed is mainly generated by the base flows in the spring and winter and by precipitation in the summer and spring.

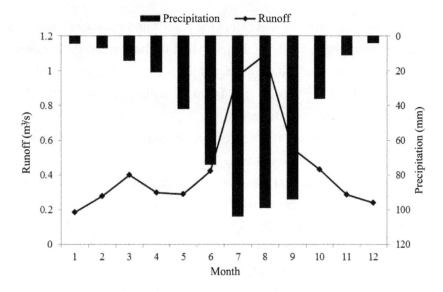

Figure 3. Comparison of the monthly measured runoff and precipitation.

Figure 4 shows the distributions of the monthly mean precipitation and sediment yields between 2001 and 2010. The annual sediment yield is generally consistent with the runoff pattern. The sediment yield is mainly concentrated in flood seasons. The maximum sediment output (16,406 t/m) occurs in August. Both the maximum runoff and the maximum sediment yield occur one month after the maximum precipitation. In the Loess Plateau, sediment production is usually accompanied by runoff with eroding capability. That the maximum peaks of runoff and sediment yield occurred within the same month is understandable. Besides precipitation, runoff and sediment generation processes are related to many other factors, including soil properties, topography, and land cover change. These factors are heavily influenced by human activities. In this watershed, a series of infrastructure reforms, water and soil conservation, and ecological projects, especially the well-known Grain-to-Green Program started in 1998 [48], were undertaken before and during the study period. Specific practices, including changing slopes into terrace fields, afforestation in barren land, constructing silt dams and reservoirs, returning farmland to forest or grassland *etc.*, could retard the processes of runoff generation and soil erosion [49].

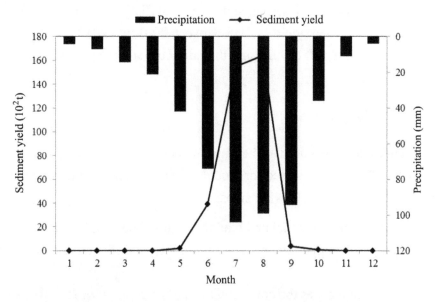

Figure 4. Comparison of the monthly measured sediment yield and precipitation.

In general, high-intensity precipitation will affect the runoff and sediment yield in a basin only after time has passed since the precipitation. The sediment yield in the watershed is mainly concentrated in the summer because high-intensity precipitation frequently occurs in that season, and precipitation is the main source of the runoff, which is main driving force of sediment output; therefore, the runoff and sediment exhibit significant seasonal variations [9].

Inter-annual runoff changes are often affected by factors such as climate change, human activities, and changes in the underlying surface conditions. Table 2 shows the statistical characteristics of the precipitation and runoff at ZHS. The annual mean precipitation and runoff at the ZHS from 2001 to 2010 are 513.38 mm and 169.04×10^6 m^3, respectively. The maximum precipitation at ZHS (634.3 mm) occurred in 2007, and the maximum runoff (235.89×10^6 m^3) occurred in 2002. The minimum precipitation (441.6 mm) and minimum runoff (95.08×10^6 m^3) at ZHS both occurred in 2008. The Cv values of the inter-annual precipitation and runoff between 2001 and 2010 are both relatively small (0.13 and 0.29, respectively). Therefore, in this 10-year period, the variation in precipitation in the watershed was insignificant, and the water resources were relatively stable, favoring the use of water resources.

Table 2. Annual characteristics of precipitation, runoff and sediment yield during 2001–2010.

Variable	Mean Value	Cv	Maximum Value		Minimum Value	
			Value	Year of Occurrence	Value	Year of Occurrence
Precipitation	513.38 mm	0.13	634.30 mm	2007	441.60 mm	2008
Runoff	169.04×10^6 m^3	0.29	235.89×10^6 m^3	2002	95.08×10^6 m^3	2008
Sediment	$1{,}330.20 \times 10^4$ t	0.84	$3{,}161.40 \times 10^4$ t	2002	122.70×10^4 t	2006

The sediment discharge is related to many factors, such as the topography and landforms of the basin, vegetation cover, precipitation, and precipitation intensity. In recent decades, the changes in the sediment yield in the Xichuan Watershed were relatively complicated due to the impacts of climate change and human activities. The mean sediment yield at ZHS between 2001 and 2010 is 1330.2×10^4 t (Table 2). During 2001 to 2010, the sediment discharge changed significantly, and soil erosion was relatively severe. The Cv value of the inter-annual sediment discharge at ZHS between 2001 and 2010 is relatively large (0.84). The maximum sediment yield at ZHS between 2001 and 2010 (3161.4×10^4 t) occurred in the same year as the maximum runoff (2002), and the minimum sediment yield (122.7×10^4 t) occurred in 2006.

3.2. SWAT Calibration and Validation Results

In this study, the year of 2001 was used as the warming period, data in 2002–2006 was used for model calibration, and data in 2007–2010 was used for model validation. The calibration procedure follows the method introduced by [36] with the following steps:

Step 1: model initiation—run the model and read the required data and parameters;

Step 2: runoff simulation—produce the simulated monthly runoffs and compare them to the actual observed values;

Step 3: if runoff simulation reaches the condition of $-20\% < Re < 20\%$, $R^2 > 0.6$, and $Ens > 0.5$, then go to the next step for sediment yield simulation; otherwise, the adjust the parameters "Base flow recession constant," "Snow pack temperature lag factor," "Soil evaporation compensation factor," "Available water capacity," "Threshold depth of water in the shallow aquifer required for return flow to occur," and "Groundwater 'revap' coefficient" and go back to step 2;

Step 4: sediment yield simulation—produce the simulated monthly sediment yields and compare them to the actual observed values;

Step 5: if sediment yield simulation reaches the condition of $-20\% < Re < 20\%$, $R^2 > 0.6$, and $Ens > 0.5$, then the simulation is successfully ended; otherwise, adjust the parameters "USLE equation support practice factor," "Linear parameters for calculating the channel sediment rooting," and "Peak rate adjustment factor for sediment routing in the main channel" and go back to step 4.

Figures 5 and 6 show comparisons between the observed and simulated monthly runoffs and sediment yield at ZHS during the model calibration period. Figures 7 and 8 show the results during the model validation period. Table 3 compares the parameters of the simulated monthly runoff and sediment yield in the model calibration and validation periods.

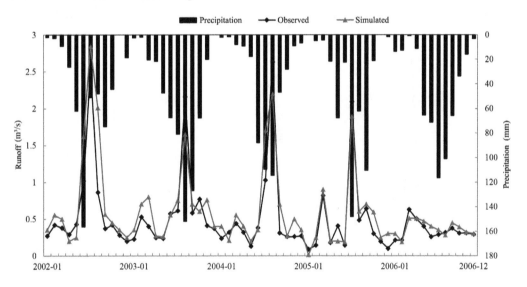

Figure 5. Comparison between the simulated and observed monthly runoff in model calibration.

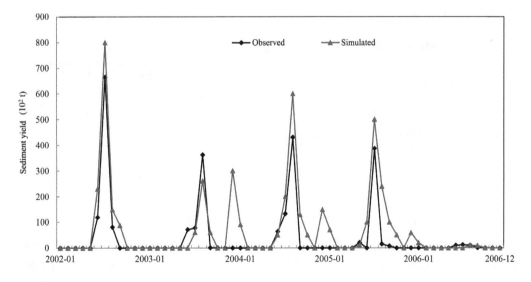

Figure 6. Comparison between the simulated and observed monthly sediment yield in model calibration.

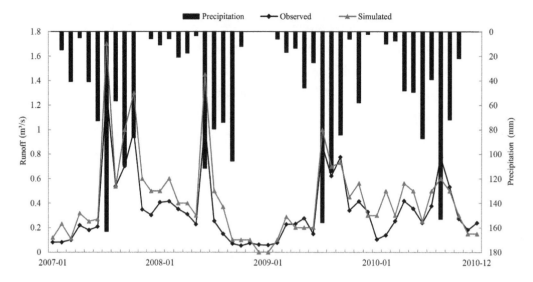

Figure 7. Comparison between the simulated and observed monthly runoff in model validation.

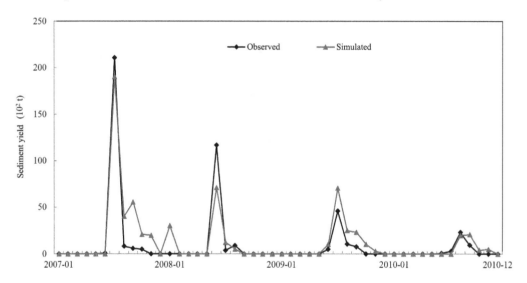

Figure 8. Comparison between the simulated and observed monthly sediment yield in model validation.

Table 3. Evaluation results of the SWAT model performance.

Simulation Period	Runoff			Sediment Yield		
	Re	R^2	*Ens*	*Re*	R^2	*Ens*
Calibration (2002–2006)	9.10%	0.79	0.73	14.20%	0.78	0.67
Verification (2007–2010)	11.20%	0.88	0.82	17.50%	0.83	0.71

As illustrated in Figures 5–8, the difference between simulated runoff/sediment yield and observed values is smaller during the calibration period than during the validation period, while the variation trend of runoff/sediment yield is more consistent with the trend of observed data during the validation period than in the calibration period. These differences can also be supported by the statistics in Table 3. Table 3 also shows that in the model validation period, the values of *Re* between the simulated and observed monthly runoffs and between the simulated and observed monthly sediment yields are 11.2% and 17.5%, respectively, and the values of R^2 are 0.88 and 0.83, respectively. The *Ens* values of runoffs and sediment yields are 0.82 and 0.71, respectively. The SWAT simulated values generally reflect the actual changes in the runoff and sediment yield, and the SWAT model can be used for the subsequent scenario analysis.

3.3. Responses of Runoff and Sediment Yield to Precipitation Changes

We also simulated the runoff and sediment yields under the four precipitation scenarios described in Section 2.4. Keeping the other inputs the same, the four precipitation scenarios were input to the validated SWAT model, and the daily runoff and sediment yield were simulated for the year of 2002 to 2010. Table 4 shows the nine-year (2002–2010) mean values. Figure 9 shows the trends and changes in the runoff and sediment yield under these precipitation scenarios.

Table 4. Responses of annual runoff and sediment yield to precipitation changes.

Simulated Item	Compared Value	P	P (1% + 20%)	P (1% + 10%)	P (1% − 10%)	P (1% − 20%)
Runoff	Simulated value (m³)	156.14	207.20	184.81	135.28	112.58
Sediment yield	Simulated value (10⁴ t)	101.09	120.49	112.87	90.93	85.04

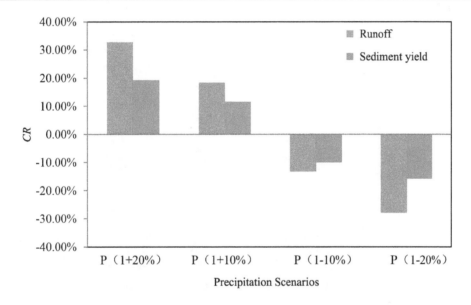

Figure 9. Changing rate of annual runoff and sediment yield change to precipitation change.

The results show that the runoff and sediment yield in this watershed under the four precipitation scenarios have the following characteristics:

(1) The runoff and sediment yield increase with increasing precipitation and decrease with decreasing precipitation, which is consistent with the actual situation. Precipitation has a direct impact on runoff, and sediment is transported by runoff. Therefore, the trends of the changes in the precipitation, runoff, and sediment are similar.

(2) When precipitation increases by 10%, the runoff and sediment yield increase by 18.36% and 11.54%, respectively. When precipitation decreases by 10%, the runoff and sediment yield decrease by 13.36% and 10.05%, respectively. The increases in the runoff and sediment yield are greater than the decreases in the runoff and sediment yield. The change in the runoff with precipitation is greater than the change in the sediment yield with precipitation. Therefore,

precipitation has a more significant impact on the runoff than the sediment yield. The runoff generated by precipitation is only one of several factors that affect sediment production and sediment production may also be affected by other factors, such as vegetation cover, soil bulk density and land use changes.

(3) When precipitation increases by 20%, the runoff and sediment yield increase by 32.7% and 19.20%, respectively. Thus, water resources will become relatively abundant when the annual precipitation intensity is relatively high, so it will be necessary to focus on preventing floods and sediment loss. When precipitation decreases by 20%, the runoff and sediment yield decrease by 27.9% and 15.88%, respectively. In these cases, water resources will be relatively scarce and it is necessary to take measures to prevent and combat droughts.

Based on the simulation results, the Cv values of the annual runoff and sediment yield are statistically calculated for the four precipitation scenarios. Figure 10 shows the trends and changes in the Cv values of the annual runoff and sediment yield for the precipitation scenarios.

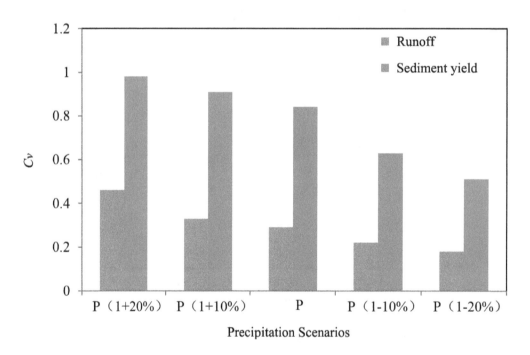

Figure 10. Coefficients of variation of annual runoff and sediment yield to precipitation changes.

The Cv values of the annual runoff and sediment yield both decrease with decreasing precipitation. The Cv values vary between 0.18 and 0.46 for the annual runoff, and vary between 0.51 and 0.98 for the sediment yield. Therefore, the Cv values of the annual runoff are smaller than those of the sediment yield. The decreased Cv value of the sediment yield with decreasing precipitation is more apparent than its increase with increasing precipitation, indicating that the Cv of the sediment yield is more sensitive to a decrease in precipitation. When precipitation increases by 20%, the Cv value of the annual runoff is 0.46 and that of the annual sediment yield is 0.98, which also demonstrates that when the precipitation intensity is relatively high, the annual runoff and sediment yield fluctuate significantly, and floods and soil erosion will occur frequently, which will be detrimental to the utilization and management of the water resources in the watershed.

4. Discussion

By studying the impacts of different precipitation scenarios on runoff and sediment yield in the Xichuan Watershed, we found that the runoff and sediment yield in the basin increase with increasing precipitation and decrease with decreasing precipitation and that the increase is more significant than the decrease. Precipitation has a more significant impact on runoff than on sediment yield. However, researchers have not yet reached a consensus about whether runoff and sediment production is more sensitive to a decrease or an increase in precipitation, as well as whether runoff or sediment production is more sensitive to changes in precipitation.

In a study of the response of runoff production in the Fox Basin in Illinois (US) to changes in precipitation, Elias [19] found that runoff production is more sensitive to increases in precipitation. In an investigation of the impact of climate change on runoff and sediment production in the purple, hilly area of Sichuan Province, China, Zeng et al. [50] found that the annual mean runoff and sediment yield increased with increasing precipitation. For the same change in precipitation, the percentage change in the sediment yield was almost twice the percentage change in the runoff. The runoff and sediment yield were more sensitive to decreases in precipitation than to increases, and the changes in the runoff and sediment yield with decreasing precipitation were more significant. The surface conditions of an area can thus affect the impact of precipitation on the runoff and sediment production in the area. In addition, another important parameter of climate change, namely temperature, will affect runoff variations. In the parameter sensitive analysis, the snow pack temperature factor is the second most sensitive parameter in the SMAT model [36], which means that melting snow is an important source of precipitation in the study area. Xia et al. [51] also discovered that when the temperature varied between 0 °C and 1 °C, the monthly runoff in the Hanjiang River Basin increased with increasing precipitation more than it decreased with decreasing precipitation. However, with increasing temperature, the extent to which the monthly mean runoff increased with increasing precipitation gradually decreased, whereas the extent to which it decreased with decreasing precipitation gradually increased. This might be attributed to the fact that an increase in temperature had a greater impact than precipitation on evapotranspiration in the basin.

This study uses the SWAT model to simulate changes in the runoff and sediment yield under several precipitation scenarios and preliminarily reveals the characteristics of the changes in the runoff and sediment yield in a small basin typical of the hill and gully area of the Loess Plateau. However, several problems in this study merit further research and investigation. (1) Several climate factors have impacts on runoff and sediment production. However, this study only considers the impact of precipitation and does not consider the impact of changes in temperature on the runoff and sediment production. Evapotranspiration is an important factor that affects surface runoff. Due to the lack of evapotranspiration data for the basin, it is not possible to calibrate and verify the evapotranspiration in the basin. Therefore, this study does not investigate the changes in evapotranspiration in the basin that are caused by changes in precipitation. Additional studies should be conducted to investigate the impact of other factors (e.g., temperature, evapotranspiration) on the runoff and sediment production; (2) This study did not consider the spatial distribution of precipitation that will bring uncertainty to the results of model simulation. Because the study area only covers 719 km^2, the assumption of a uniform precipitation is acceptable. When the method is extended to large river basins, the spatial distribution of precipitation should be considered. Since the HRU is the basic spatial unit in the simulation, the spatial distribution of precipitation can be considered by inputting different precipitation data for different HRUs if more precipitation stations are available. (3) Due to the data availability, the precipitation scenarios were set using the empirical method, a method used by other researchers in this situation [21,23]. In future studies, efforts should be made to collect basic data and use climate output models to predict precipitation to conduct in-depth investigations on the trends of future climate change.

5. Conclusions

A SWAT model was parameterized in the Xichuan Watershed, a typical hilly and gully loess area in the Loess Plateau, China. The variations of runoff and sediment yield were simulated using the calibrated SWAT model and scenario analyses. Based on the simulation, we found that the increases of runoff and sediment yield with increasing precipitation are more apparent than their decreases with decreasing precipitation. Precipitation has a more significant impact on runoff than on sediment yield. The Cv values of the annual runoff and sediment yield both increase with increasing precipitation. However, the Cv value of the annual runoff is relatively smaller than the Cv value of the sediment yield when precipitation increases, and the Cv value of sediment yield is more sensitive to a decrease in precipitation than is that of runoff. The different characteristics of variations in runoff and sediment yield suggests proper strategies in the utilization and management of the water resources in this watershed.

Acknowledgments

This work was supported by the National Natural Science Foundation of China with Grant No. 50979003. Support from the Collaborative Innovation Center for Regional Environmental Quality is also acknowledged. Special thanks also extend to the editors and all of the anonymous reviewers.

Author Contributions

Tianhong Li conceived and designed the research, and wrote the paper; Yuan Gao analyzed the data, localized SWAT model and performed scenario analysis.

References

1.   Allen, M.R.; Ingram, W.J. Constrains on future changes in climate and the hydrologic cycle. *Nature* **2002**, *419*, 224–232.
2.   Oki, T.; Kanae, S. Global hydrological cycles and world water resources. *Science* **2006**, *313*, 1068–1072.
3.   Dong, L.H.; Xiong, L.H.; Yu, K.X.; Li, S. Research advances in effects of climate change and human activities on hydrology. *Adv. Water Sci.* **2012**, *2*, 278–285. (In Chinese)
4.   Trenberth, K.E. Conceptual framework for changes of extremes of the hydrological cycle with climate change. In *Weather and Climate Extremes—Changes, Variations and a Perspective from the Insurance Industry*; Springer Netherlands: Dordrecht, The Netherlands, 1999; pp.327–339.

5.   Zhang, J.Y.; Wang, G.Q. *Research on Impacts of Climate Change on Hydrology and Water Resources*; Science Press: Beijing, China, 2007. (In Chinese)

6.   Piao, S.L.; Ciais, P.; Huang, Y.; Shen, Z.; Peng, S.; Li, J.; Zhou, L.; Liu, H.; Ma, Y.; Ding, Y.; *et al.* The impacts of climate change on water resources and agriculture in China. *Nature* **2010**, *467*, 43–51.

7.   Barnett, T.P.; Adam, J.C.; Lettenmaier, D.P. Potential impacts of a warming climate on water availability in snow-dominated regions. *Nature* **2005**, *438*, 303–309.

8.   IPCC. *Climate Change 2007: Impacts, Adaptation and Vulnerability Contribution of Working Group 11 to the Fourth Assessment Report of the Intergovernmental Panel on Climate Change*; Cambridge University: Cambridge, UK; New York, NY, USA, 2007.

9.   Li, F.P.; Zhang, G.X.; Dong, L.Q. Studies for impact of climate change on hydrology and water resources. *Sci. Geogr. Sin.* **2013**, *4*, 457–464.

10. Song, X.M.; Zhang, J.Y.; Zhan, C.S.; Liu, C.Z. Review for impacts of climate change and human activities on water cycle. *China J. Hydrol.* **2013**, *44*, 779–790. (In Chinese)

11. Xia, J.; Liu, C.Q.; Ren, G.Y. Opportunity and challenge of the climate change impact on the water resources of China. *Adv. Earth Sci.* **2011**, *1*, 1–12.

12. Wang, G.Q.; Zhang, J.Y.; Liu, J.F.; Jin, J.L.; Liu, C.S. The sensitivity of runoff to climate change in different climatic regions in China. *Adv. Water Sci.* **2011**, *3*, 307–314. (In Chinese)

13. Arnold, J.G.; Williams, J.R.; Srinivasan, R.; King, K.W. *The Soil and Water Assessment Tool (SWAT) User's Manual*; USDA-ARS: Temple, TX, USA, 1995.

14. Eckhardt, K.; Ulbrich, U. Potential impacts of climate change on groundwater recharge and streamflow in a central European low mountain range. *J. Hydrol.* **2003**, *284*, 244–252.

15. Ficklin, D.L.; Luo, Y.Z.; Luedeling, E.; Zhang, M. Climate change sensitivity assessment of a highly agricultural watershed using SWAT. *J. Hydrol.* **2009**, *374*, 16–29.

16. Xu, Z.X.; Zuo, D.P.; Tang, F.F. *Response of Water Cycle to Future Climate Change in Typical Watershed in the Yellow River Basin*; Press of Hehai University: Nanjing, China, 2012; pp. 37–49. (In Chinese)

17. Githui, F.; Gitau, W.; Mutuab, F.; Bauwens, W. Climate change impact on SWAT simulated streamflow in Western Kenya. *Int. J. Climatol.* **2009**, *29*, 1823–1834.

18. Yu, L.; Gu, J.; Li, J.X.; Zhu, X.J. A Study of hydrologic responses to climate change in medium scale basin based on SWAT. *Bull. Water Soil Conserv.* **2008**, *28*, 152–154. (In Chinese)

19. Bekele, E.G.; Knapp, H.V. Watershed Modeling to Assessing Impacts of Potential Climate Change on Water Supply Availability. *Water Resour. Manag.* **2010**, *24*, 3299–3320.

20. Zhu, C.H.; Li, Y.K. Long-term hydrological impacts of land use/land cover change from 1984 to 2010 in the Little River Watershed, Tennessee. *Int. Soil Water Conserv. Res.* **2014**, *2*, 11–22.

21. Nowak, K.; Hoerling, M.; Rajagopalan, B.; Zagona, E. Colorado River Basin Hydro-climatic Variability. *J. Clim.* **2012**, *25*, 4389–4403.

22. Fan, Y.T.; Chen, Y.N.; Li, W.H.; Wang, H.J.; Li, X.G. Impacts of temperature and precipitation on runoff in the Tarim River during the past 50 years. *J. Arid Land* **2011**, *3*, 220–230.

23. Cayan, D.R.; Dettinger, M.D.; Kammerdiener, S.A.; Caprio, J.M.; Peterson, D.H. Changes in the onset of spring in the western United States. *Bull. Am. Meteorol. Soc.* **2001**, *82*, 399–416.

24. Colman, R. A comparison of climate feedbacks in general circulation models. *Clim. Dynam.* **2003**, *20*, 865–873.

25. Liu, Q.; Yang, Z.; Cui, B. Spatial and temporal variability of annual precipitation during 1961–2006 in Yellow River Basin, China. *J. Hydrol.* **2008**, *361*, 330–338.

26. Lu, X.X. Vulnerability of water discharge of large Chinese rivers to environmental changes: An overview. *Reg. Environ. Chang.* **2004**, *4*, 182–191.

27. Wei, W.; Chen, L.; Fu, B.; Chen, J. Water erosion response to rainfall and land use in different drought-level years in a loess hilly area of China. *Catena* **2010**, *81*, 24–31.

28. Miao, C.Y.; Ni, J.R.; Borthwick, A.G.L.; Yang, L. A preliminary estimate of human and natural contributions to the changes in water discharge and sediment load in the Yellow River. *Glob. Plenary Chang.* **2011**, *76*, 196–205.

29. Miao, C.Y.; Ni, J.R.; Borthwick, A.G.L. Recent changes of water discharge and sediment load in the Yellow River basin, China. *Phys. Geogr.* **2010**, *34*, 541–561.

30. Ren, Z.P.; Zhang, G.H.; Yang, Q.K. Characteristics of runoff and sediment variation in the Yanhe River Basin in last 50 years. *J. China Hydrol.* **2012**, *32*, 81–86. (In Chinese)

31. Sui, J.; He, Y.; Liu, C. Changes in sediment transport in the Kuye River in the Loess Plateau in China. *Int. J. Sediment Res.* **2009**, *24*, 201–213.

32. Li, Z.; Liu, W.Z.; Zhang, X.C.; Zheng, F.L. Impacts of land use change and climate variability on hydrology in an agricultural catchment on the Loess Plateau of China. *J. Hydrol.* **2009**, *377*, 35–42.

33. Luo, H.M.; Li, T.H.; Ni, J.R.; Wang, Y.D. Water Demand for Ecosystem Protection in River with Hyper-concentrated Sediment-laden Flow. *Sci. China Ser. E* **2004**, *47*, 186–198.

34. Saxton, K.E. Soil Water Characteristics Hydraulic Properties Calculator [EB/OL]. Available online: http: //www.bsyse.wsu.edu/saxton/soilwater (accessed on 20 March 2004).

35. Saxton, K.E.; Willey, P.H.; Rawls, W.J. Field and pond hydrologic analyses with the SPAW model. In Proceedings of the Annual International Meeting of American Society of Agricultural and Biological Engineers, Portland, OR, USA, 9–12 July 2006; pp.1–13.

36. Gao, Y.; Li, T.H. Responses of runoff and sediment yield to LUCC with SWAT model: A case study in the Xichuan River Basin, China. *Sustain. Environ. Res.* **2015**, *25*, 27–35.

37. Bai, Y.M. Effect of returning farmland to forests to benefit of soil and water conservation in Xichuanhe River basin. *J. Water Resour. Water Eng.* **2011**, *22*, 176–178. (In Chinese)

38. Liu, K.W. River Systems in Yanan City; Press of Yan'An Education College: Yan'an, China, 2000. (In Chinese)

39. USDA (United States Department of Agriculture). *Urban Hydrology for Small Watersheds: TR-55*; No. 210-VI-TR-55; Government Printing Office: Washington, DC, USA, 1986.

40. Williams, J.; Nearing, M.; Nicks, A.; Skidmore, E.; Valentin, C.; King, K.; Savabi, R. Using soil erosion models for global change studies. *J. Soil Water Conserv.* **1996**, *51*, 381–385.

41. Chiew, F.H.S.; McMahon, T.A. The applicability of Morton's and Penman's evapotranspiration estimates in rainfall-runoff modelling. *Water Resour. Bull.* **1991**, *27*, 611–620.

42. Williams, J.R. Flood routing with variable travel time or variable storage coefficients. *Trans. ASAE* **1969**, *12*, 100–103.

43. Van Griensven, A.; Meixner, T.; Grunwald, S.; Bishop, T.; Diluzio, M.; Srinivasan, R. A global sensitivity analysis tool for the parameters of multi-variable catchment models. *J. Hydrol.* **2006**, *324*, 10–23.

44. Nagelkerke, N.J.D. A Note on a General Definition of the Coefficient of determination. *Biometrika* **1991**, *78*, 691–692.

45. Quinlan, J.R. Learning with continuous classes. In Proceedings of the AI'92, the 5th Australian Joint Conference on Artificial Intelligence, Tasmania, Australia, 16–18 November 1992; Adams, A., Sterling, L., Eds.; World Scientific: Singapore, 1992; pp. 343–348.

46. Nash, J.E.; Sutcliffe, J.V. River flow forecasting through conceptual models. Part I—A discussion of principles. *J. Hydrol.* **1970**, *10*, 282–290.

47. Yue, B.J.; Shi, Z.H.; Fang, N.F. Evaluation of rainfall erosivity and its temporal variation in the Yanhe River catchment of the Chinese Loess Plateau. *Nat. Hazards* **2014**, *74*, 585–602.

48. Fu, B.J.; Wang, Y.F.; Lu, Y.H.; He, C.H.; Chen, L.D.; Song, C.J. The effects of land-use combinations on soil erosion: A case study in the Loess Plateau of China. *Prog. Phys. Geogr.* **2009**, *33*, 793–804.

49. Zhu, L.Q.; Zhu, W.B. Research on effects of land use/cover change on soil erosion. *Adv. Mater. Res.* **2012**, *433*, 1038–1043.

50. Zeng, Y.; Wei, L. Impacts of Climate and Land Use Changes on Runoff and Sediment Yield in Sichuan Purple Hilly Area. *Bull. Water Soil Conserv.* **2013**, *33*, 1–6. (In Chinese)

51. Xia, Z.H.; Zhou, Y.H.; Xu, H.M. Water resources responses to climate changes in Hanjiang River basin based on SWAT model. *Resour. Environ. Yangtze Basin* **2010**, *19*, 158–163. (In Chinese)

6

Tracking Inflows in Lake Wivenhoe during a Major Flood using Optical Spectroscopy

Rupak Aryal [1,*], Alistair Grinham [2] and Simon Beecham [1]

[1] Centre for Water Management and Reuse, School of Natural and Built Environments,
University of South Australia, Mawson Lakes, SA 5095, Australia;
E-Mail: Simon.Beecham@unisa.edu.au

[2] School of Civil Engineering, The University of Queensland, St Lucia, QLD 4072, Australia;
E-Mail: a.grinham@uq.edu.au

* Author to whom correspondence should be addressed; E-Mail: rupak.aryal@unisa.edu.au;

Abstract: Lake Wivenhoe is the largest water storage reservoir in South-East Queensland and is the primary drinking water supply storage for over 600,000 people. The dam is dual purpose and was also designed to minimize flooding downstream in the city of Brisbane. In early January, 2011, record inflows were experienced, and during this period, a large number of catchment pollutants entered the lake and rapidly changed the water quality, both spatially and vertically. Due to the dendritic nature of the storage, as well as multiple inflow points, it was likely that pollutant loads differed greatly depending on the water depth and location within the storage. The aim of this study was to better understand this variability in catchment loading, as well as water quality changes during the flood event. Water samples were collected at five locations during the flood period at three different depths (surface, mid-depth and bottom), and the samples were analysed using UV and fluorescence spectroscopy. Primary inflows were identified to persist into the mid-storage zone; however, a strong lateral inflow signature was identified from the mid-storage zone, which persisted to the dam wall outflow. These results illustrate the heterogeneity of inflows in water storages of this type, and this paper discusses the implication this has for the modelling and management of such events.

Keywords: lake water quality; flooding; optical spectroscopy

1. Introduction

Lake Wivenhoe, situated 80 km west of Brisbane, is one of the largest dams in Australia. The lake has a capacity to store 1.15 million megalitres (ML) of water and is the major water supply to Brisbane, which is the fourth largest city in Australia. Being situated on the banks of the Brisbane River, the city frequently experiences flooding. Wivenhoe Dam lies on this river, approximately 80 km upstream of the city of Brisbane. It was specifically designed to minimize the flood risk to Brisbane. During flood periods, the lake is capable of holding back a total of 1.45 million ML. After a long decade of drought, Brisbane experienced extreme rainfall between the end of December, 2010, and the first week of January, 2011. The resultant runoff rapidly filled Lake Wivenhoe to 190% of its designed storage capacity. The surrounding catchment is heavily modified with only 40% remnant vegetation, and this, combined with the record inflows, resulted in a significant pollutant loading, including sediment, dissolved organic matter and nutrients, over a very short period of time [1].

Dissolved organic matter (DOM) is of great concern, due to its role in the binding of nutrients, heavy meals and other pollutants from surrounding terrestrial environments. DOM influences the physical and chemical environment in lakes through light attenuation and metal complexion [2–4]. DOM is also important in trophic dynamics [5,6], which promote the growth of heterotrophic microorganisms [7,8]. Furthermore, the incorporation of nitrogen and phosphorus into the DOM pool can influence nutrient cycling in lakes and reservoirs [9–12]. DOM can negatively impact water treatment directly through taste, odour and colour issues and during chlorination through the production of disinfection by-products. Finally, DOM can lead to bacterial proliferation within water distribution systems.

DOM comprises a large number of organic molecules of varied composition, and their characterization can be both complicated and labour intensive. However, monitoring the spatial and vertical variation of DOM is useful for gaining a better understanding of aquatic environmental significance, particularly during periods of major catchment inflows. During major inflows, water quality can change significantly in short periods of time, and simple and sensitive tools are required to rapidly provide qualitative information regarding DOM changes. Optical spectroscopy techniques, such as UV and fluorescence spectroscopy, are both rapid and capable of providing useful characterization of a wide range of DOM.

Optical spectroscopies, such as UV and fluorescence spectroscopy, have been extensively used to characterize organic matter that undergoes changes due to chemical, biological and physical processes in water and wastewater. Optical spectroscopies are popular, because they are reasonably sensitive, simple, rapid and economic. The UV technique can be used with absorption on single- or dual-wavelength procedures and can provide information on individual or representative organic chemical species. The specific wavelength can provide information on numerous chemicals present in the environmental samples [13–17]. The fluorescence spectroscopy method, commonly known as the excitation emission matrix (EEM), is a technique that can be used to obtain an optical fingerprint of dissolved organic matter in water and wastewater, and this can provide information on the nature of microbial, humic and fulvic organics and other pollutants, such as hydrocarbons. Its high sensitivity and its specificity to specific chemicals or groups of chemicals, such as amino acids, aromatic amino acids, mycosporine-like amino acids, humics, proteins and fulvic type substances, have made the application of fluorescence widely popular in the last few years in environmental monitoring [18–20]. Although UV and EEM have been used to monitor water and wastewater in the past, their application in tracking DOM and specific chemical constituents both spatially and vertically has not been reported so far.

The main aim of this paper is to demonstrate how simple optical techniques can be used to track DOM inflows in a lake during flood periods where rapid mixing of water both spatially and vertically takes place.

2. Materials and Methods

Sampling was conducted on 21 January 2011, 10 days after the peak inflows occurred. In order to maximize the spatial representation of the lake, sites were selected from the dam wall, through the main body of the lake and at adjacent major inflow points (Figure 1). Water samples were taken from both the surface (20 cm below the water surface), mid (8 m water depth) and bottom (>15 m, 1 m above the sediment surface) with a vertical, 4.2-L Niskin water sampler (Wildco, Wildlife Supply Company, Yulee, FL, USA). Water depth at each site was recorded from an on-board depth sounder (Lowrance Elite-7 HDI, Navico Inc., Ensenada, MX, USA). Prior to sample collection, the Niskin and sampling bottles (glass) were cleaned with diluted nitric acid followed by Milli-Q water and twice flushed with water from the same sample depth to minimise contamination. Field personnel took care to not handle the inside of the Niskin or sample containers during sampling, and samples were placed on ice after collection for transport to the laboratory.

Figure 1. Lake Wivenhoe and sampling stations.

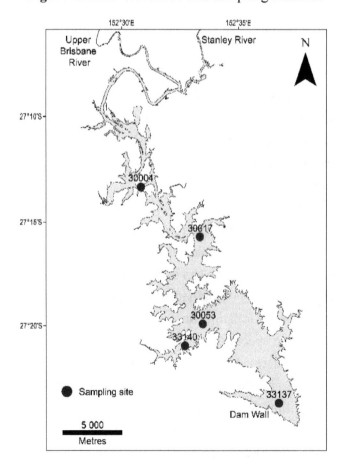

The laboratory samples were filtered through a 1.2-μm filter (Whatman GF/C, GE Healthcare, Little Chalfont, UK) to avoid the influence of turbidity due to suspended solids that cause light scattering, shading and, thus, influence the absorption over the entire spectrum. The filtrate was analysed for dissolved organic carbon (DOC), UV and fluorescence spectra. Details are described below.

2.1. Dissolved Organic Carbon Analysis

Dissolved organic carbon was measured by liquid chromatography with online organic carbon detection (LCOCD, Karlsruhe, Germany) [21]. No replicates were performed in this study.

2.2. UV Analysis

The water samples were analysed using a UV spectrometer (Varian 50 Bio, Victoria, Australia). The instrument was operated at a bandwidth of 1 nm, with a quartz cell of a 10-mm path length, a wavelength of 190 to 400 nm and at a scanning speed of 190 nm/min (slow) at room temperature, 22 ± 2 °C. Milli-Q water was recorded as blank at every set of experiments and subtracted from each sample's UV record.

2.3. Fluorescence Analysis

Three-dimensional fluorescence spectra, also known as excitation emission matrix (EEM) spectra, were obtained using a spectrofluorometer (Perkin Elmer LS 55, Victoria, Australia) with a wavelength range of 200 nm to 500 nm (for excitation); and 280 nm to 500 nm (for emission). The spectra were taken at an incremental wavelength of 5 nm in excitation (Ex); and 2 nm in emission (Em). The EEM value of blank (MQ water) data was subtracted from the analysed samples for blank correction. The fluorescence intensity was corrected by blank subtraction and was expressed in quinine sulphate units (QSU) [22].

A 290-nm emission cut-off filter was used to eliminate the second order Rayleigh light scattering. To eliminate water Raman scatter peaks, Milli-Q water was recorded as the blank and subtracted from each sample. The inner filter effect of EEMs caused by possible higher concentrations of dissolved organic matter (DOM) in the samples was corrected for absorbance by the multiplication of each value in the EEM with a correction factor based on the idea that the average path length of the absorption of the excitation and emission light is 1/2 the cuvette length. For this purpose, the expression was used:

$$F_{corr} = F_{obs} \times 10^{(\lambda_{ex} + \lambda_{em})/2} \tag{1}$$

where F_{corr} and F_{obs} are the corrected and observed fluorescence intensities and λ_{ex} and λ_{em} are the absorbances at the current excitation and emission wavelengths.

The data obtained from EEM were analysed using an "R" program according to Chen *et al.* (2003) [23] and described below.

$$\emptyset_i = \sum_{ex} \sum_{em} (\lambda_{ex}\lambda_{em}) \Delta\lambda_{ex}\lambda_{em} \tag{2}$$

where $\emptyset_i$ is the volume beneath region i; $(\lambda_{ex}\lambda_{em})$ is the fluorescence intensity at each excitation-emission wavelength pair and $\Delta\lambda_{ex}$ and $\Delta\lambda_{em}$ are the excitation and emission wavelength intervals, respectively.

$$\emptyset_T = \sum \emptyset_i \qquad (3)$$

where $\emptyset_T$ is the cumulative volume.

The EEM spectra was divided into five major regions (shown in Table 1 and Figure 2). The program could calculate area, as well as the contribution percentage of each area.

Table 1. Five major regions in excitation emission matrix (EEM) spectra according to *Chen et al.* (2003) [23]. SMP, soluble microbial by-product; FA, fulvic acid; HA, humic acid. (P1 and P2 = proteins, Ex = excitation, Em = emission, and BOD = biological oxygen demand.)

Region	Chemical composition of organic matter
I (P1): Ex:Em 200–250:280–330	lower molecular weight tyrosine-like aromatic amino acids
II (P2): Ex:Em 200–250:330–380	low molecular weight aromatic proteins and BOD-type substances
III (SMP): Ex:Em 250–340:280–380	large molecular weight peptides and proteins (microorganism related by-products)
IV (FA): Ex:Em 200–250:380–500	fulvic acid type substances
V (HA): Ex:Em 250–500:380–500	humic acid type substances

Figure 2. Five EEM regions selected for this study from the surface water of Site 33137 (regions plotted according to Chen *et al.*, 2003) [23].

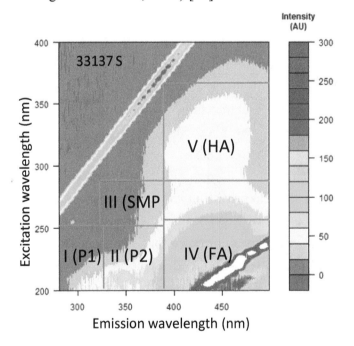

3. Results and Discussion

3.1. Spatial and Vertical Variation in DOM Concentration

Table 2 shows dissolved organic carbon (DOC) across sampling sites and vertically within each site. The DOC concentration was 3–4-times lower than at stratified conditions (9.6–12.8 mg/L recorded almost one year after this study). The lower DOC concentrations indicate dilution by flooding waters. The DOC decreased from inflow sites in the upper lake downstream to the lower lake. Surface waters of upstream sites had relatively elevated DOC levels compared to deeper waters. However, at the middle and downstream sites, the bottom waters had elevated DOC relative to mid-depth and surface waters. Higher DOC concentrations adjacent to a major inflow point (Site 30004) was assumed to be due to catchment inflows having relatively less dilution with lake water compared with sites further into the lake. Decreases in the surface DOC at sites further into the lake were possibly due to settling, microbial/photochemical decomposition and/or subsurface catchment inflows tracking through the lake at deeper depths compared with shallower inflow points [24]. A higher rate of degradation of humic substances at the surface when exposed to UV is reported by Salonen and Vahatalo in a lake in Findland [25].

Table 2. Distribution of dissolved organic carbon (mg·L^{-1}) and turbidity (nephelometric turbidity units—NTU) (in brackets) distribution spatially and vertically in Lake Wivenhoe post flood period.

Site (location)	Surface	Mid-depth	Bottom
30004 (upstream)	2.782 (29.8)	2.498 (40.7)	2.622 (191.2)
30017 (upstream)	2.624 (68.9)	2.312 (67.7)	2.248 (212.8)
30053 (middle)	2.129 (77.4)	3.433 (128.2)	2.843 (229)
33140 (middle)	2.244 (90.6)	1.748 (107.8)	2.218 (175.1)
33137 (downstream)	2.373 (111.3)	2.188 (116.5)	2.929 (210.6)

3.2. Optical Analysis

In both UV and fluorescence spectroscopy, incident radiation causes the loosely bound electron present in double or triple bonds and/or in electronegative elements to excite. The absorption of incident radiation is recorded in UV spectroscopy against the wavelength according to the Beer–Lambert Law ($A = \log \left(I_o / I \right)$), where I_o is the incident radiation and I is the radiation after passing through the length of solution. In fluorescence spectroscopy, the energy released by excited species to come to the ground state is also recorded. The specific excitation and emission wavelengths are unique for particular species. Two molecules may have similar excitation energies, but different emission energies.

3.2.1. UV Spectra

UV spectroscopy is rapid, simple and requires little sample preparation and small volume samples. Within the absorbance range between 190 and 400 nm, many specific absorbance values are related to a variety of properties, such as aromaticity, hydrophobic content, apparent molecular weight and size and biodegradability [26–28]. Table 3 summarises popular wavelengths widely used to measure chemical species in water and wastewater.

Figure 3 shows a contour diagram of the UV spectral intensity of water recorded at various wavelengths (195, 215, 254 and 330 nm) in Lake Wivenhoe samples collected at the surface (Figure 3a), mid-depth (Figure 3b) and bottom (Figure 3c) during the flood period in January, 2011. Colour patterns in the contour diagram reflect the absorbance intensity at particular wavelengths.

Table 3. UV absorbance recorded at various wavelengths used to measure chemical species in water and wastewater. COD = chemical oxygen demand.

Wavelength (nm)	Property	Reference
195	Proteins	[29]
210	Amino acids	[14,30]
215	Peptides	[30,31]
230	Proteins	[32]
254	Aromaticity	[33]
260	Hydrophobic content/COD	[16,34]
265	Relative abundance of functional group	[35]
272	Aromaticity	[36]
280	Hydrophobic carbon index	[37]
285	Humification index	[27]
300	Characterisation of humic substances	[38]
310–360	Mycosporine-like amino acids	[39–41]
350	Apparent molecular size	[15]
365	Aromaticity, apparent molecular weight	[42]

For wavelengths of 195 nm (proteins), 215 nm (amino acids) and 254 nm (aromaticity), similar colour patterns in the contour diagram indicated that the DOM in the surface and bottom waters showed more homogeneity than in the mid-depth water. In the mid-depth region, UV 215 nm and UV 254 nm showed two distinct colour patterns separating the middle regions from the upstream inflows (surface) and downstream outflows (bottom). The middle region of the lake, where the inflows align, has similar organic characteristics, but at the edge of the lake, different organics are evident. Two possible reasons are proposed: turbid flood runoff being stored in the middle of the reservoir and preferential pathways of stormwater passing through the lake when flowing from upstream to downstream. This is supported by higher DOC and higher turbidity in the middle of the lake at Site 30053. Similar results (higher DOC and turbidity) have been observed by Kim *et al.* (2000) in a deep reservoir at Lake Soyang, Korea [43] when stormwater flooded into the reservoir.

The UV 330 nm wavelength represents mycosporine-like amino acids (MAAs) [39,40]. MAAs are small colourless water soluble compounds composed of cyclohexane or cyclohexenimine chromophore conjugated with nitrogen substituents of amino acids or imino alcohol [44,45] and are very susceptible to photodegradation [41,46]. On the surface, various colour bands were observed in the contour diagram from the upstream inflows to the downstream outflows, but these were not evident in the mid-depth and bottom regions. The results indicate that for the inflows, photodegradation occurred over time in the surface region.

Figure 3. Contour diagram of the UV spectral intensity of flood water recorded at wavelengths 195, 215, 254 and 330 nm collected at the surface (**a**), mid-depth (**b**) and bottom (**c**) of Lake Wivenhoe (colours indicate the absorbance intensity).

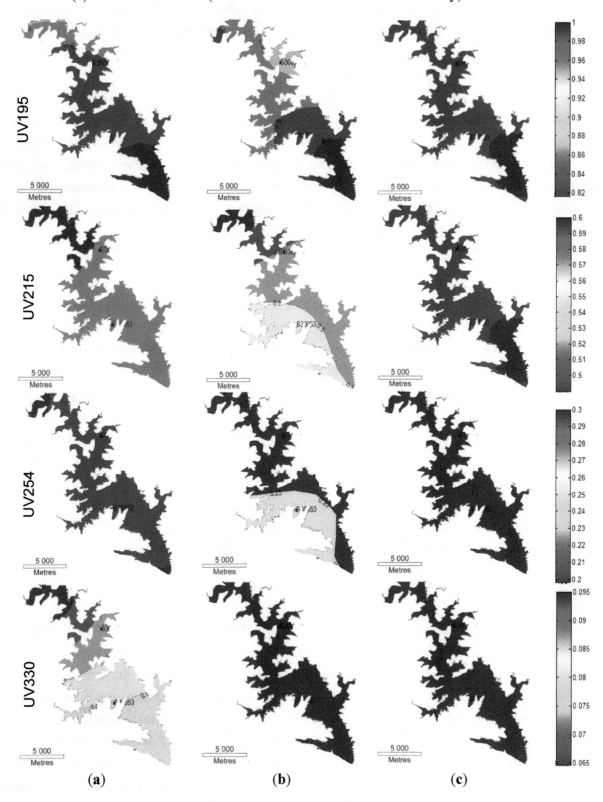

3.2.2. Fluorescence Spectra

Unlike UV, fluorescence spectra provide both excitation and emission information simultaneously, which can distinguish two chemical species from each other that have similar excitation energies. The fluorescence spectra provide information on different types of organics [47]. There are a number of different methods for interpreting fluorescence data, from selective peak picking [48,49] to complex modelling, such as parallel factor analysis [50,51]. One of the commonly adopted methods is to calculate the area of peak of specific region of the spectrum proposed by (Chen *et al.*, 2003) [23]. In this paper, the EEM spectrum was divided into five regions, and the area of the peak was calculated using "R" software, following the equations described previously [23]. According to Chen *et al.* (2003) [23], Regions I (P1) and II (P2) represent aromatic proteins, Region III represents soluble microbial by-product-like (SMP) substances, Region IV represents fulvic acid-like (FA) substances and Region V represents humic acid-like (HA) substances. Table 1 summarises the representative chemicals in the EEM spectra in the five excitation:emission regions selected for this study.

Figure 4 shows a contour diagram of the five chemical species P1 to HA at three depths. The colour patterns represent fluorescence intensity (or relative concentration) of specific groups of chemicals in the lake, as shown in the legend (far right). Among the five groups of chemicals measured on the surface, P1, P2 and SMP seem to be influenced by the flow pattern, as shown by a preferential flow route for these species. In contrast, the spillover of FA and HA were not affected by the flow. At the mid-depth, we observed a relative decrease in the concentration of P1, P2 and SMP with flow from upstream, but this is not the case with FA and HA. This result indicates the removal of P1, P2 and SMP by sedimentation (binding with particles) and/or by chemical conversion to other organics. We also observed higher fluorescence intensities for P1, P2 and SMP substances in the middle region of the bottom of the lake. The increased microbial activity observed in the middle region of the surface of the water shows possible stagnation of the flow in this region. At the bottom of the lake, sediment particle settling provides increased surface area for microbial colonization, thus allowing increased rates of activity in these waters. According to Kim *et al.* (2000) [43] , the carbon and nutrients held in the middle region of turbid water become a good source of food for various bacteria.

4. Conclusions

Lake Wivenhoe is the primary supply water for the city of Brisbane, and it is essential to understand the impact of flood events on water quality in order to fully understand catchment loadings into the system, as well as possible implications for water treatment. The DOC concentration varied spatially and vertically, indicating that the inflow of DOM in the lake varied with space, as well as depth. The UV and fluorescence spectral techniques used in this study showed that organic species were distributed heterogeneously across the lake, both spatially and vertically, and information on specific chemicals or groups of chemicals can be obtained easily. These findings demonstrate the feasibility of optical spectroscopy techniques for understanding the impacts of catchment inflows on DOM species across the lake, and the findings will be highly beneficial for both water treatment and asset management.

Figure 4. Contour diagrams of the fluorescence spectral intensity of flood water for the surface, mid-depth and bottom of water across the lake based on the fluorescence area of each region (colours represent fluorescence intensity).

Acknowledgments

The authors wish to thank Seqwater for logistical and financial support and, in particular, Deb Gale and Cameron Veal for their assistance in sample collection.

Author Contributions

The experimental design of the project was undertaken by Rupak Aryal and Alistair Grinham. The majority of field work was conducted by Alistair Grinham. Rupak Aryal, conducted chemical analysis with the help of coauthors Alistair Grinhama and Simon Beecham. All three authors involved equally in interpreting the data and writing this manuscript.

References

1. Grinham, A.; Gibbes, B.; Gale, D.; Watkinson, A.; Bartkow, M. Extreme rainfall and drinking water quality: A regional perspective. *Proc. Water Pollut.* **2012**, *164*, 183–194.

2. Zafiriou, O.C.; Joussot-Dubien, J.; Zepp, R.G.; Zika, R.G. Photochemistry of natural waters. *Environ. Sci. Technol.* **1984**, *18*, 358A–371A.

3. Mostofa, K.M.; Wu, F.; Liu, C.-Q.; Vione, D.; Yoshioka, T.; Sakugawa, H.; Tanoue, E. Photochemical, microbial and metal complexation behavior of fluorescent dissolved organic matter in the aquatic environments. *Geochem. J.* **2011**, *45*, 235–254.

4. Davis, J.A. Complexation of trace metals by adsorbed natural organic matter. *Geochim. Cosmochim. Acta* **1984**, *48*, 679–691.

5. Tranvik, L.; Kokalj, S. Decreased biodegradability of algal DOC due to interactive effects of UV radiation and humic matter. *Aquat. Microb. Ecol.* **1998**, *14*, 301–307.

6. Jansson, M.; Bergström, A.-K.; Blomqvist, P.; Drakare, S. Allochthonous organic carbon and phytoplankton/bacterioplankton production relationships in lakes. *Ecology* **2000**, *81*, 3250–3255.

7. McKnight, D.M.; Smith, R.L.; Harnish, R.A.; Miller, C.L.; Bencala, K.E. Seasonal relationships between planktonic microorganisms and dissolved organic material in an alpine stream. *Biogeochemistry* **1993**, *21*, 39–59.

8. Stone, L.; Berman, T. Positive feedback in aquatic ecosystems: The case of the microbial loop. *Bull. Math. Biol.* **1993**, *55*, 919–936.

9. Schindler, D.; Bayley, S.; Curtis, P.; Parker, B.; Stainton, M.; Kelly, C. Natural and man-caused factors affecting the abundance and cycling of dissolved organic substances in precambrian shield lakes. *Hydrobiologia* **1992**, *229*, 1–21.

10. McKnight, D.; Thurman, E.M.; Wershaw, R.L.; Hemond, H. Biogeochemistry of Aquatic Humic Substances in Thoreau's Bog, Concord, Massachusetts. *Ecology* **1985**, *66*, 1339–1352.

11. Qualls, R.G.; Richardson, C.J. Factors controlling concentration, export, and decomposition of dissolved organic nutrients in the Everglades of Florida. *Biogeochemistry* **2003**, *62*, 197–229.

12. Mladenov, N.; McKnight, D.M.; Wolski, P.; Ramberg, L. Effects of annual flooding on dissolved organic carbon dynamics within a pristine wetland, the Okavango Delta, Botswana. *Wetlands* **2005**, *25*, 622–638.

13. GHOSH, K.; Schnitzer, M. UV and visible absorption spectroscopic investigations in relation to macromolecular characteristics of humic substances. *J. Soil Sci.* **1979**, *30*, 735–745.

14. Aitken, A.; Learmonth, M. *The Protein Protocols Handbook* 1996; Springer: New York, NY, USA, 1996; pp. 3–6.

15. Korshin, G.V.; Li, C.-W.; Benjamin, M.M. Monitoring the properties of natural organic matter through UV spectroscopy: A consistent theory. *Water Res.* **1997**, *31*, 1787–1795.

16. Dilling, J.; Kaiser, K. Estimation of the hydrophobic fraction of dissolved organic matter in water samples using UV photometry. *Water Res.* **2002**, *36*, 5037–5044.

17. Roig, B.; Thomas, O. UV spectrophotometry: A powerful tool for environmental measurement. *Manag. Environ. Qual.* **2003**, *14*, 398–404.

18. Aryal, R.; Kandel, D.; Acharya, D.; Chong, M.N.; Beecham, S. Unusual Sydney dust storm and its mineralogical and organic characteristics. *Environ. Chem.* **2012**, *9*, 537–546.

19. Hong, S.; Aryal, R.; Vigneswaran, S.; Johir, M.A.H.; Kandasamy, J. Influence of hydraulic retention time on the nature of foulant organics in a high rate membrane bioreactor. *Desalination* **2012**, *287*, 116–122.

20. Hussain, S.; van Leeuwen, J.; Chow, C.; Beecham, S.; Kamruzzaman, M.; Wang, D.; Drikas, M.; Aryal, R. Removal of organic contaminants from river and reservoir waters by three different aluminum-based metal salts: Coagulation adsorption and kinetics studies. *Chem. Eng. J.* **2013**, *225*, 394–405.

21. Huber, S.A.; Balz, A.; Abert, M.; Pronk, W. Characterisation of aquatic humic and non-humic matter with size-exclusion chromatography—Organic carbon detection—Organic nitrogen detection (LC-OCD-OND). *Water Res.* **2011**, *45*, 879–885.

22. Aryal, R.K.; Murakami, M.; Furumai, H.; Nakajima, F.; Jinadasa, H.K.P.K. Prolonged deposition of heavy metals in infiltration facilities and its possible threat to groundwater contamination. *Water Sci. Technol.* **2006**, *54*, 205–212.

23. Chen, W.; Westerhoff, P.; Leenheer, J.A.; Booksh, K. Fluorescence excitation-emission matrix regional integration to quantify spectra for dissolved organic matter. *Environ. Sci. Technol.* **2003**, *37*, 5701–5710.

24. Moran, M.A.; Sheldon, W.M., Jr.; Zepp, R.G. Carbon loss and optical property changes during long-term photochemical and biological degradation of estuarine dissolved organic matter. *Limnol. Oceanogr.* **2000**, *45*, 1254–1264.

25. Salonen, K.; Vähätalo, A. Photochemical mineralisation of dissolved organic matter in lake Skjervatjern. *Environ. Int.* **1994**, *20*, 307–312.

26. Ma, H.; Allen, H.E.; Yin, Y. Characterization of isolated fractions of dissolved organic matter from natural waters and a wastewater effluent. *Water Res.* **2001**, *35*, 985–996.

27. Kalbitz, K.; Geyer, S.; Geyer, W. A comparative characterization of dissolved organic matter by means of original aqueous samples and isolated humic substances. *Chemosphere* **2000**, *40*, 1305–1312.

28. Imai, A.; Fukushima, T.; Matsushige, K.; Kim, Y.H. Fractionation and characterization of dissolved organic matter in a shallow eutrophic lake, its inflowing rivers, and other organic matter sources. *Water Res.* **2001**, *35*, 4019–4028.

29. Yabushita, S.; Wada, K.; Inagaki, T.; Arakawa, E. UV and vacuum UV spectra of organic extract from Yamato carbonaceous chondrites. *Mon. Not. R. Astron. Soc.* **1987**, *229*, 45P–48P.

30. Aryal, R.; Vigneswaran, S.; Kandasamy, J. Application of Ultraviolet (UV) spectrophotometry in the assessment of membrane bioreactor performance for monitoring water and wastewater treatment. *Appl. Spectrosc.* **2011**, *65*, 227–232.

31. Kuipers, B.J.; Gruppen, H. Prediction of molar extinction coefficients of proteins and peptides using UV absorption of the constituent amino acids at 214 nm to enable quantitative reverse phase high-performance liquid chromatography-mass spectrometry analysis. *J. Agric. Food Chem.* **2007**, *55*, 5445–5451.

32. Liu, P.-F.; Avramova, L.V.; Park, C. Revisiting absorbance at 230 nm as a protein unfolding probe. *Anal. Biochem.* **2009**, *389*, 165–170.

33. Hur, J.; Schlautman, M.A. Using selected operational descriptors to examine the heterogeneity within a bulk humic substance. *Environ. Sci. Technol.* **2003**, *37*, 880–887.

34. Chevakidagarn, P. Surrogate parameters for rapid monitoring of contaminant removal for activated sludge treatment plants for para rubber and seafood industries in Southern Thailand. *J. Songklanakarin.* **2005**, *27*, 417–424.

35. Chen, J.; Gu, B.; LeBoeuf, E.J.; Pan, H.; Dai, S. Spectroscopic characterization of the structural and functional properties of natural organic matter fractions. *Chemosphere* **2002**, *48*, 59–68.

36. Traina, S.J.; Novak, J.; Smeck, N.E. An ultraviolet absorbance method of estimating the percent aromatic carbon content of humic acids. *J. Environ. Qual.* **1990**, *19*, 151–153.

37. Chin, Y.-P.; Aiken, G.; O'Loughlin, E. Molecular weight, polydispersity, and spectroscopic properties of aquatic humic substances. *Environ. Sci. Technol.* **1994**, *28*, 1853–1858.

38. Artinger, R.; Buckau, G.; Geyer, S.; Fritz, P.; Wolf, M.; Kim, J. Characterization of groundwater humic substances: Influence of sedimentary organic carbon. *Appl. Geochem.* **2000**, *15*, 97–116.

39. Dionisio-Sese, M.L. Aquatic microalgae as potential sources of UV-screening compounds. *Philipp. J. Sci.* **2010**, *139*, 5–16.

40. Winter, A.R.; Fish, T.A.E.; Playle, R.C.; Smith, D.S.; Curtis, P.J. Photodegradation of natural organic matter from diverse freshwater sources. *Aquat. Toxicol.* **2007**, *84*, 215–222.

41. Whitehead, K.; Vernet, M. Influence of mycosporine-like amino acids (MAAs) on UV absorption by particulate and dissolved organic matter in La Jolla Bay. *Limnol. Oceanogr.* **2000**, *45*, 1788–1796.

42. Peuravuori, J.; Pihlaja, K. Molecular size distribution and spectroscopic properties of aquatic humic substances. *Anal. Chim. Acta* **1997**, *337*, 133–149.

43. Kim, B.; Choi, K.; Kim, C.; Lee, U.H.; Kim, Y.-H. Effects of the summer monsoon on the distribution and loading of organic carbon in a deep reservoir, Lake Soyang, Korea. *Water Res.* **2000**, *34*, 3495–3504.

44. Singh, S.P.; Kumari, S.; Rastogi, R.P.; Singh, K.L.; Sinha, R.P. Mycosporine-like amino acids (MAAs): Chemical structure, biosynthesis and significance as UV-absorbing/screening compounds. *Indian J. Exp. Biol.* **2008**, *46*, 7–17.

45. Sinha, R.; Klisch, M.; Gröniger, A.; Häder, D.-P. Ultraviolet-absorbing/screening substances in cyanobacteria, phytoplankton and macroalgae. *J. Photochem. Photobiol. B* **1998**, 47, 83–94.

46. Vincent, W.F.; Roy, S. Solar ultraviolet-B radiation and aquatic primary production: Damage, protection, and recovery. *Environ. Rev.* **1993**, *1*, 1–12.

47. Chong, M.N.; Sidhu, J.; Aryal, R.; Tang, J.; Gernjak, W.; Escher, B.; Toze, S. Urban stormwater harvesting and reuse: A probe into the chemical, toxicology and microbiological contaminants in water quality. *Environ. Monit. Assess.* **2012**, 1–8.

48. Birdwell, J.E.; Engel, A.S. Characterization of dissolved organic matter in cave and spring waters using UV–Vis absorbance and fluorescence spectroscopy. *Org. Geochem.* **2010**, *41*, 270–280.

49. Coble, P.G. Characterization of marine and terrestrial DOM in seawater using excitation-emission matrix spectroscopy. *Mar. Chem.* **1996**, *51*, 325–346.

50. Stedmon, C.A.; Bro, R. Characterizing dissolved organic matter fluorescence with parallel factor analysis: A tutorial. *Limnol. Oceanogr.* **2008**, *6*, 572–579.

51. Stedmon, C.A.; Markager, S.; Bro, R. Tracing dissolved organic matter in aquatic environments using a new approach to fluorescence spectroscopy. *Mar. Chem.* **2003**, *82*, 239–254.

7

Optimal Choice of Soil Hydraulic Parameters for Simulating the Unsaturated Flow

Ken Okamoto [1], **Kazuhito Sakai** [2,*], **Shinya Nakamura** [2], **Hiroyuki Cho** [3], **Tamotsu Nakandakari** [2] **and Shota Ootani** [4]

[1]   United Graduate School of Agricultural Sciences, Kagoshima University, 1-21-24 Korimoto, Kagoshima-shi, Kagoshima 890-0065, Japan; E-Mail: r130041@eve.u-ryukyu.ac.jp

[2]   Faculty of Agriculture, University of the Ryukyus, 1 Senbaru, Nishihara-cho, Okinawa 903-0213, Japan; E-Mails: s-naka@agr.u-ryukyu.ac.jp (S.N.); zhunai@agr.u-ryukyu.ac.jp (T.N.)

[3]   Faculty of Agriculture, Saga University, 1 Honjo-machi, Saga 840-8502, Japan; E-Mail: choh@cc.saga-u.ac.jp

[4]   Eight-Japan Engineering Consultants Inc.; 33-11 Honcho 5 Chome, Nakano-ku, Tokyo 164-0012, Japan; E-Mail: ootani-sho@ej-hds.co.jp

*   Author to whom correspondence should be addressed; E-Mail: ksakai@agr.u-ryukyu.ac.jp;

Academic Editor: Athanasios Loukas

Abstract: We examined the influence of input soil hydraulic parameters on HYDRUS-1D simulations of evapotranspiration and volumetric water contents (VWCs) in the unsaturated zone of a sugarcane field on the island of Miyakojima, Japan. We first optimized the parameters for root water uptake and examined the influence of soil hydraulic parameters (water retention curve and hydraulic conductivity) on simulations of evapotranspiration. We then compared VWCs simulated using measured soil hydraulic parameters with those using pedotransfer estimates obtained with the ROSETTA software package. Our results confirm that it is important to always use soil hydraulic parameters based on measured data, if available, when simulating evapotranspiration and unsaturated water flow processes, rather than pedotransfer functions.

Keywords: HYDRUS-1D; ROSETTA; retention curve; hydraulic conductivity; sugarcane field; the island of Miyakojima

1. Introduction

There are no rivers on Miyakojima, a semitropical island in southernmost Japan, because of its flat topography and the high permeability of the limestone that forms the island. Local residents there are dependent on groundwater for almost all of their domestic water use. In the Okinawa region, temperatures are predicted to rise in response to climate change, while annual rainfall is expected to decrease [1], with resultant depletion of groundwater resources becoming a concern. The main land use on Miyakojima is sugarcane farming; thus, understanding both water movement in the unsaturated zone of the farmland soil and the total water budget is important.

The HYDRUS-1D software package [2] has often been used for analyses of these type of problems [3,4]. HYDRUS-1D provides versatile numerical modeling of the movement of moisture, solutes, and heat in soil. One option in the code is to estimate soil hydraulic properties by using pedo-transfer functions (PTFs). Since it is difficult to measure soil hydraulic parameters, PTFs that estimate them from readily measurable soil characteristics, such as particle size distribution and bulk density provide a very attractive tool for numerical analyses.

The accuracy of HYDRUS-1D simulations has been analytically verified [5]. Most studies to evaluate the performance of HYDRUS-1D were simulations of the transfer of heat and moisture in semiarid and humid regions. For example, Saito *et al.* [6] reported that HYDRUS-1D was useful for predicting the transfer of heat and moisture in sandy soils, and Kato *et al.* [7] reported that it was useful for predicting soil temperature and moisture in volcanic soils.

Other studies have used vadose zone models and PTFs. Steinzer *et al.* [8] simulated evapotranspiration, infiltration, and VWCs distribution in lysimeter experiments on sandy and clay soils in Germany. Wang *et al.* [9] simulated groundwater recharge in sand and sandy loam soil by using HYDRUS-1D. They found that recharge in their soils was strongly dependent on the parameter *n* of the van Genuchten model [10] and uncertainties in the simulated recharge were affected by uncertainties in the *n* estimated from PTFs.

Thus, although HYDRUS-1D is known to be useful for simulations of water movement in the unsaturated zone, the code must be tested in a particular area before practical applications; for example, for the development of a water management plan. Wang *et al.* [11] simulated groundwater recharge at four sites in the continental United States with different climate conditions using HYDRUS-1D along with datasets for sand and loamy sand. They showed that the distribution patterns of mean annual groundwater recharge varied considerably across the sites, mainly depending on soil texture and climatic conditions.

To date, the use of HYDRUS-1D in the island of Miyakojima has not yet been tested. Although the necessary weather data are readily available from the Japan Meteorological Agency, the collection of soil data is more difficult. Most of the soil data collected and analyzed in past investigations for land development projects consists only of particle size distributions and bulk density; water retention curves were not always determined. Accurate simulation results are dependent on the quality of the soil hydraulic parameters used as input to the simulation. Okamoto *et al.* [12] reported that the retention curve of Shimajiri mahji soil (dark-red soil [13], which was classified as a Cambisol [14]) estimated using the ROSETTA software package [15–17] was considerably different from that derived from measured data. However, they did not comment on the influence of this difference on the simulated water budget.

To validate the use of HYDRUS-1D on Miyakojima, we examined the influence of input soil hydraulic parameters on HYDRUS-1D simulations of evapotranspiration and VWCs. We first optimized the parameters for root water uptake and examined the influence of soil parameters on simulations of evapotranspiration (details are shown in the later section). We then compared VWCs simulated using measured soil hydraulic parameters with those simulated using parameters derived by using ROSETTA software.

2. Materials and Methods

2.1. Study Site

Our study site is in a sugarcane field at Saratake in the Shirakawada groundwater basin, which is one of several fault-bounded groundwater basins on Miyakojima (Figure 1). We measured precipitation (CTK-15PC, Climatec, Inc., Tokyo, Japan), temperature (CVS-HMP-155D, Climatec, Inc., Tokyo, Japan), wind speed (CPR010C, Climatec, Inc., Tokyo, Japan), net radiation (CHF-NR01, Climatec, Inc., Tokyo, Japan), evapotranspiration (CS7500, Campbell Scientific, Logan, UT, USA and SAT-540, SONIC Co., Tokyo, Japan), and VWCs (EC5, Decagon Devices, Pullman, WA, USA). Evapotranspiration was measured by the eddy covariance method [18]. Time-domain reflectometry soil moisture sensors were installed at depths of 15, 30, 50, and 70 cm. Data were collected every 30 min. Measurement started on 29 August 2009 and ended on 31 December 2009. Cultivation started on 21 February 2009 and ended on 20 January 2010.

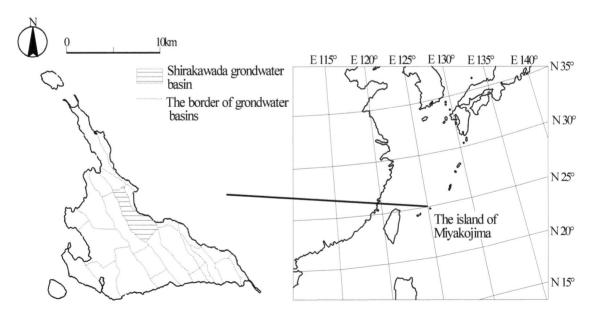

Figure 1. Groundwater basins on the island of Miyakojima.

The surface soil at the study site is a Jahgaru soil (gray soil [13,19]), which extends to a depth of 120 cm at our sampling site. Jahgaru soil derived from Shimajiri-mudstone is classified as Calcaric Regosols [14]. We sampled both disturbed and undisturbed soil at depths of 15, 30, 50, 70, and 100 cm. Undisturbed soil samples were collected using a 100 cm^3 soil core sampler (inside diameter 5 cm, height 5.1 cm).

We measured soil particle size distributions (%), bulk densities (g·cm^{-3}), hydraulic conductivities (cm·day^{-1}), and water retention curves in the laboratory. Soil particle size distributions were obtained using sieve analysis for particle sizes greater than 75 μm and by hydrometer analysis for particle sizes smaller than 75 μm, according to Japanese Industrial Standards (JIS A1202). We measured bulk densities by using the 100 cm^3 soil core sampler, hydraulic conductivities by the constant-head method, and water retention curves by the three methods shown in Table 1.

Table 1. Experimental procedures used for VWC measurements at selected soil suction ranges.

Method	Log$_{10}$ [Suction (cm)]								
	1.0	1.5	2.0	2.5	3.0	3.5	4.0	4.5	5.0
Hanging water column	————————————								
Pressure plate				————————					
Psychrometer *							————————————		

Note: * WP4-T, Decagon Devices, Pullman, WA, USA.

2.2. HYDRUS-1D

2.2.1. Overview of HYDRUS-1D

One-dimensional water flow in soil is described by the Richards equation as follows:

$$\frac{\partial \theta}{\partial t} = \frac{\partial}{\partial z}\left(K \frac{\partial h}{\partial z} \right) + \frac{\partial K}{\partial z} - S, \tag{1}$$

where θ is volumetric water content (cm^3·cm^{-3}), t is time (s), z is the spatial coordinate, assumed positive upward (cm), h is pressure head (cm), S is a sink term for water uptake by plant roots (cm^3·cm^{-3}·s^{-1}), and K is unsaturated hydraulic conductivity (cm·s^{-1}). We applied the van Genuchten-Mualem equation [10,20] (VG hereafter) to estimate the water retention curve and hydraulic conductivity of the soil by using Equations (2) and (3), respectively:

$$S_e = \frac{\theta - \theta_r}{\theta_s - \theta_r} = \left(1 + |\alpha h|^n\right)^{-m}, \tag{2}$$

$$K(h) = K_s S_e^{\,l}\left[1 - S_e\left(1 - S_e^{1/m}\right)^m\right]^2, \tag{3}$$

where θ_r is residual water content (cm^3·cm^{-3}), θ_s is saturated water content (cm^3·cm^{-3}), and S_e is normalized water content. α (cm^{-1}), n, m ($= 1 - 1/n$) and l ($= 0.5$) are empirical parameters, and K_s is saturated hydraulic conductivity (cm s^{-1}).

We simulated five soil layers (0–20, 20–40, 40–60, 60–80, and 80–120 cm) that correspond to the depth ranges of the soil samples we collected. Atmospheric conditions of daily precipitation and potential evaporation were used as upper boundary conditions. Potential transpiration was used to calculate water uptake by plant roots. The lower boundary condition was free drainage. Initial VWC was the VWC measured in the field.

2.2.2. Crop Model in HYDRUS-1D

We used the method of van Genuchten et al. [21] to simulate water uptake by plant roots:

$$S(h) = \alpha(h) S_p, \tag{4}$$

$$\alpha(h) = \frac{1}{1 + \left(\dfrac{h}{h_{50}} \right)^p}, \tag{5}$$

where $\alpha(h)$ is the water stress response function, S_p is potential water uptake by plants roots, h_{50} is the soil suction at which water uptake by roots is reduced by 50%, and p is an empirical component that is usually assumed to be 3 [22]. We used $p = 3$ to optimize h_{50}.

To determine daily root length, we used a logistic root growth function in HYDRUS-1D [2] and the results of a previous study [23]. We used the leaf area index (LAI) growth model of Larsbo and Jarvis [24]. The LAI growth model from crop emergence day (D_{min}) to the day of maximum LAI (D_{max}) is described by Equation (6), and that from D_{max} to the day of harvest (D_{harv}) by Equation (7):

$$LAI = LAI_{min} + \left(LAI_{max} - LAI_{min} \right) \left(\frac{D^* - D_{min}}{D_{max} - D_{min}} \right)^{x_1}, \tag{6}$$

$$LAI = LAI_{harv} + \left(LAI_{max} - LAI_{harv} \right) \left(\frac{D_{harv} - D^*}{D_{harv} - D_{max}} \right)^{x_2}, \tag{7}$$

where LAI_{min} is LAI at D_{min}, LAI_{max} is LAI at D_{max}, LAI_{harv} is LAI at D_{harv}, D^* is the number of days after planting, and x_1 and x_2 are empirical components. We derived daily values of LAI from the results of a previous study of sugarcane [25] (Figure 2).

Potential evapotranspiration (ET_p) was calculated using the Penman equation (8) [26,27]:

$$ET_p = \frac{\Delta}{\Delta + \gamma} \cdot \frac{S}{L} + \frac{\gamma}{\Delta + \gamma} \cdot u_2 \left(e_{sa} - e_a \right), \tag{8}$$

where ET_p is the reference evapotranspiration (mm·day^{-1}), Δ is the slope of the saturation vapor pressure curve (kPa·°C^{-1}), γ is the psychometric constant (kPa·°C^{-1}), S is the net radiation (MJ·m^{-2}·day^{-1}), L is the latent heat of evaporation (MJ·kg^{-1}) is, u_2 is wind speed at 2 m height (m·s^{-1}), ($e_{sa} - e_a$) is the saturation vapor pressure deficit (kPa). ET_p was partitioned into potential evaporation (E_p) and potential transpiration (T_p) using Campbell's equation [28]:

$$T_p = ET_p \left[1 - \exp(-8.2 LAI) \right] \tag{9}$$

$$E_p = ET_p - T_p \tag{10}$$

Total precipitation during the observation period was 675.5 mm (Figure 3). Since the days after planting during the observation period varied from 189 to 313 and *LAI* from 2.4 to 3.3 (from Equations (6) and (7)), E_p was low throughout the observation period.

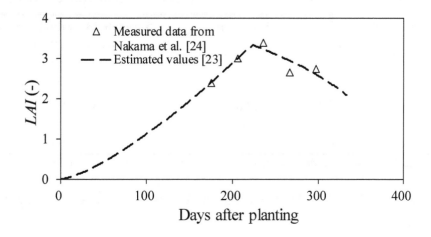

Figure 2. Change of *LAI* after planting of sugarcane.

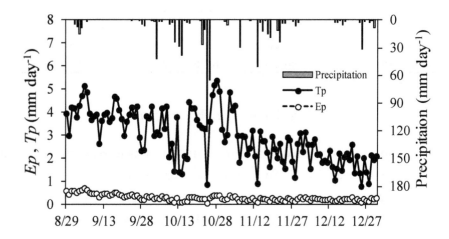

Figure 3. Daily precipitation, potential evaporation (*Ep*) and potential transpiration (*Tp*) from 29 August to 31 December 2009.

2.3. Comparison of Measured Soil Hydraulic Parameters with Those Estimated Using ROSETTA

We determined the parameters for the VG equation by using our analyses of soil samples and by application of the ROSETTA module in HYDRUS-1D. We determined VG parameters so as to minimize the difference between measured values and estimated values according to the nonlinear least-squares method using the solver of an EXCEL add-in. ROSETTA can use combinations of soil texture, particle size distribution, bulk density, and one or two points on the water retention curve to determine the parameters for the VG equation and for hydraulic conductivity. In this study, we used particle size distribution and bulk density, since these properties were readily available.

Measured particle size distribution, bulk density and hydraulic conductivity are shown in Table 2. For all samples, the sand content was less than 20% and the soil texture was silt. There were no clear differences in bulk density among soil layers. The standard deviation of hydraulic conductivity was large for all layers, which we attributed to cracks formed in response to shrinkage during drying of the soil samples. We, therefore, used the geometric mean of hydraulic conductivity in the simulations.

Table 2. Measured particle distribution, bulk density, and hydraulic conductivity of all layers.

z (cm)	Sand (%)	Silt (%)	Clay (%)	Bulk Density (g·cm^{-3})	Saturated Conductivity, K_s Geometric mean ± SD (cm·day^{-1})
15	13.1	69.6	17.3	1.319	24.7 ± 97.8
30	11.1	71.2	17.7	1.278	24.3 ± 152.6
50	7.5	75.3	17.2	1.268	21.8 ± 90.0
70	8.7	74.8	16.5	1.305	116.3 ± 343.8
100	10.4	70.3	19.4	1.159	21.9 ± 99.8

Note: Particle size distribution was derived from one sample and hydraulic conductivity from four samples.

In our application of HYDRUS-1D, we considered four combinations of input soil hydraulic parameters (Cases 1 to 4; Table 3).

Table 3. Combinations of parameters input to HYDRUS-1D.

Soil Hydraulic Parameter	Case 1	Case 2	Case 3	Case 4
Retention curve	Measured	ROSETTA	ROSETTA	Measured
Hydraulic conductivity	Measured	ROSETTA	Measured	ROSETTA

2.4. Simulation of Evapotranspiration and Volumetric Water Contents

To optimize the value of h_{50}, we needed to minimize the difference between measured and simulated evapotranspiration for each of Cases 1 to 4. To achieve this, for each case we ran HYDRUS-1D using h_{50} values from 100 to 1000 cm at intervals of 100 cm. Then, we chose the optimum value of h_{50} as the value with the lowest root-mean-square error (RMSE) for total evapotranspiration calculated at 10-day intervals.

To examine the influence of input soil hydraulic parameters on simulated VWCs movement, we calculated the RMSE between measured VWCs and simulated VWCs (optimized h_{50}) for each of the four soil layers for Cases 1 to 4.

3. Results and Discussion

3.1. Comparison of Measured Soil Hydraulic Parameters with Those Estimated by ROSETTA

In general, ROSETTA underestimated measured VWCs, particularly for soil suctions greater than 1000 cm (Figure 4). Okamoto *et al.* [12] reported similar results from their application of ROSETTA to Shimajiri mahji soil. Since Jahgaru soil has a poorly developed structure [19] it often does not drain well [18]. Therefore, the measured VWC values at high suction tended to be larger than that estimated by ROSETTA (Figure 4). Schaap and Leij [29] reported that the use of PTFs to estimate soil hydraulic parameters might depend strongly on the data used for calibration. The characteristics of Jahgaru soil might be different from the soil data used to develop on the parameters used in ROSETTA.

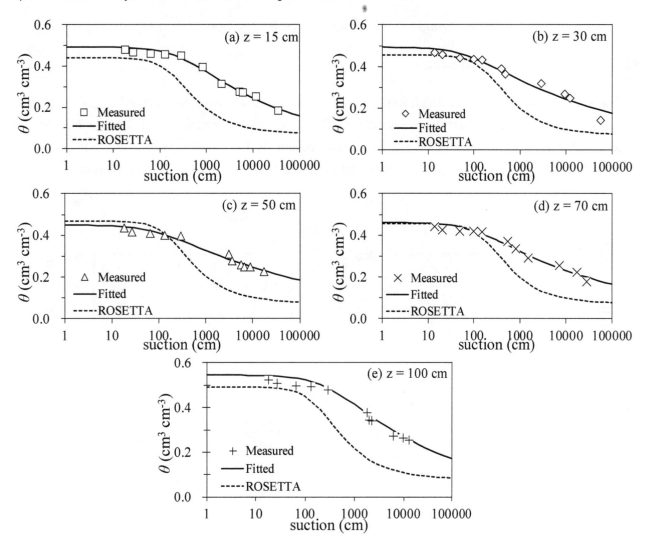

Figure 4. Soil water retention curves from measured data and as estimated by ROSETTA. (**a**) z = 15 cm; (**b**) z = 30 cm; (**c**) z = 50 cm; (**d**) z = 70 cm; (**e**) z = 100 cm.

Measured hydraulic conductivities were 21.8–116.3 cm d^{-1}, whereas those estimated by ROSETTA were 30.3–53.5 cm·d^{-1} (Table 4). We attributed the difference in these results to the influence of cracks and aggregations of soil in the sugarcane field.

Table 4. Parameters used in simulations.

z (cm)	θ_r (cm^3·cm^{-3})		θ_s (cm^3·cm^{-3})		α (cm^{-1})		n		K_s (cm·d^{-1})	
	Fitted	ROS	Fitted	ROS	Fitted	ROS	Fitted	ROS	Measured	ROS
15	0	0.070	0.493	0.441	0.0037	0.0049	1.193	1.685	24.7	30.3
30	0	0.073	0.493	0.456	0.0148	0.0049	1.140	1.683	24.3	35.0
50	0	0.074	0.449	0.469	0.0117	0.0051	1.125	1.672	21.8	35.2
70	0	0.072	0.462	0.456	0.0107	0.0050	1.147	1.677	116.3	32.3
100	0	0.078	0.556	0.492	0.0082	0.0049	1.166	1.675	21.9	53.5

Note: ROS = from ROSETTA.

3.2. Simulation of Evapotranspiration and Volumetric Water Contents

3.2.1. Optimization of h_{50} and Simulation of Evapotranspiration

The simulations run to optimize h_{50} (Figure 5) show that total evapotranspiration (TET) increased with increasing h_{50} for all four cases considered. The optimal h_{50} (smallest RMSE) was 600 cm for Case 1, 400 cm for Case 2, 300 cm for Case 3 and 700 cm for Case 4. For each of the four cases, the TET estimated with the optimized h_{50} was almost the same as the measured TET (204.5 mm).

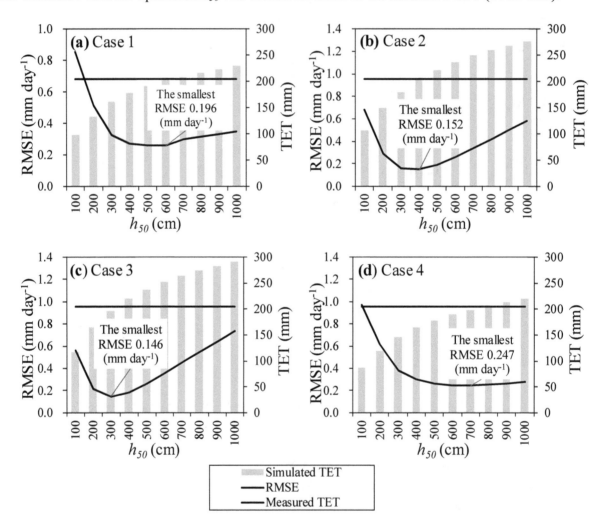

Figure 5. Relationships of the value of h_{50} to RMSE and TET for each Case. (**a**) Case1; (**b**) Case 2; (**c**) Case 3; (**d**) Case 4.

The simulation results for the pairs of cases with the same retention curve (Cases 1 and 4, Cases 2 and 3) were similar (compare Figures 5 and 6). The suction required to deplete VWCs during normal growth has been reported to be about 1000 cm [30]; therefore, we considered that Case 1 (600 cm) and Case 4 (700 cm) provided the more realistic values of h_{50} for application in HYDRUS-1D, even though the RMSEs of Cases 2 and 3 were lower. The common factor for Cases 1 and 4 was the use of the measured retention curve.

Thus, comparison of our simulation results for evapotranspiration did not clearly indicate which soil hydraulic parameters were better for application in HYDRUS-1D.

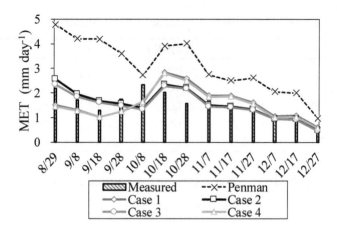

Figure 6. Comparison of simulated, measured and potential evapotranspiration at ten-day intervals (MET) from 29 August to 31 December 2009.

3.2.2. Influence of Soil Hydraulic Parameters on Volumetric Water Contents Simulation

The time series of simulated VWCs (Figure 7) and RMSEs between measured and simulated VWCs (Table 5) show that the simulated VWC values for Cases 2 and 3 (retention parameters estimated using ROSETTA) were lower than measured values, whereas for Cases 1 and 4 (retention parameters calculated from measured data) simulated and measured VWCs agreed well.

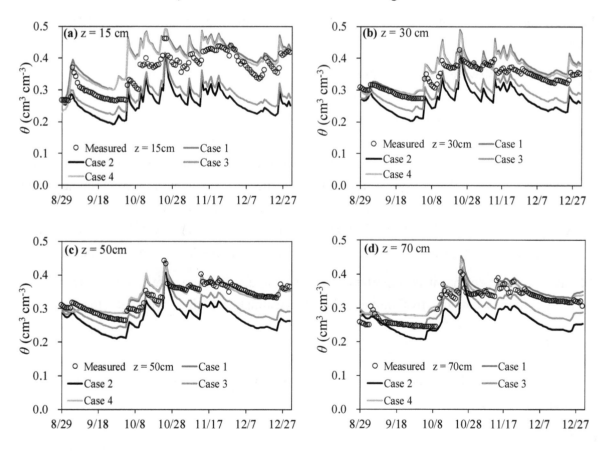

Figure 7. Comparison of simulated and measured daily VWCsat 4 depths from 29 August to 31 December 2009. **(a)** z = 15 cm; **(b)** z = 30 cm; **(c)** z = 50 cm; **(d)** z = 70 cm.

Table 5. RMSE between measured and simulated VWCs.

z (cm)	RMSE (cm³·cm⁻³)			
	Case 1	Case 2	Case 3	Case 4
15	0.052	0.101	0.083	0.043
30	0.035	0.067	0.047	0.028
50	0.017	0.060	0.040	0.017
70	0.026	0.051	0.031	0.021

When considering the water budget, water recharge (Re) is calculated using

$$R_e = P - ET - \Delta SW , \tag{11}$$

where P is precipitation, ET is evapotranspiration and ΔSW is the change of VWCs. To estimate water recharge from this equation, accurate simulations of VWCs and evapotranspiration are needed. The results of our simulations of evapotranspiration for Cases 1 to 4 (Figure 6) did not differ greatly (Section 3.2.1). However, our simulations of VWCs indicated that Cases 1 and 4 provided better results than Cases 2 and 3. These results indicate the importance of using soil hydraulic parameters based on retention curves derived from measured data to calculate the water budget for the island of Miyakojima, and likely for many or most other applications.

4. Conclusions

We drew the following main conclusions about the optimum application of HYDRUS-1D in our study area.

1) ROSETTA software underestimated measured VWCs.
2) Optimized values of h_{50} were dependent on the parameters defined by the retention curve. Simulated and measured total evapotranspiration rates agreed well for all four cases considered. Since, for normal growth the amount of suction required to deplete VWCs is about 1000 cm, we consider the h_{50} values we obtained that were closest to 1000 cm to be the more realistic. Thus, our HYDRUS-1D simulations using the measured soil hydraulic parameters provided better results than those based on parameters estimated by ROSETTA.
3) VWCs simulated by HYDRUS-1D using parameters estimated by ROSETTA were lower than the measured values, whereas those using measured parameters agreed well with measured values.

Our study confirmed that it is important to use soil hydraulic parameters derived from measured retention data on the island of Miyakojima, rather than estimates obtained with pedotransfer function.

Acknowledgements

The authors gratefully acknowledges the constructive comments and suggestions from the anonymous reviewers.

Author Contributions

Ken Okamoto designed and performed the experiments, analyzed and interpreted the data and wrote the manuscript. Kazuhito Sakai designed the study and interpreted the results. Shinya Nakamura,

Hiroyuki Cho, Tamotsu Nakandakari and Shota Ootani interpreted the data. All authors read and approved the final manuscript.

References

1. Climate Change and Its Impacts in Japan. Available online: https://www.env.go.jp/en/earth/cc/report_impacts.pdf (accessed on 4 September 2015).

2. Ficklin, D.L.; Luedeling, E.; Zhang, M. Sensitivity of groundwater recharge under irrigated agriculture to changes in climate, CO_2 concentrations and canopy structure. *Agric. Water Manag.* **2010**, *97*, doi:10.1016/j.agwat.2010.02.009.

3. Leterme, B.; Mallants, D.; Jacques, D. Sensitivity of groundwater recharge using climatic analogues and HYDRUS-1D. *Hydrol. Earth Syst. Sci.* **2012**, *16*, doi:10.5194/hess-16-2485-2012.

4. Šimůnek, J.; Šejna M.; Saito, H.; van Genuchten, M.T. *The HYDRUS-1D Software Package for Simulating the One-Dimensional Movement of Water, Heat, and Multiple Solutes in Variably-Saturated Media*; Department of Environmental Science, University of California Riverside: Riverside, CA, USA, 2013.

5. Zlotnik, V.A.; Wang, T.; Nieber, J.L.; Šimunek, J. Verification of numerical solutions of the Richards equation using a traveling wave solution. *Adv. Water Resour.* **2007**, *30*, 1973–1980.

6. Saito, H.; Šimůnek, J.; Mohanty, B.P. Numerical analysis of coupled water, vapor, and heat transport in the vadose zone. *Vadose Zone J.* **2006**, *5*, doi:10.2136/vzj2006.0007.

7. Kato, C.; Nishimura, T.; Imoto, H.; Miyazaki, T. Predicting soil moisture and temperature of andisoils under a monsoon climate in Japan. *Vadose Zone J.* **2011**, *10*, 541–551.

8. Stenitzer, E.; Diestel, H.; Zenker, T.; Schwartengräber, R. Assessment of capillary rise from shallow groundwater by the simulation model SIMWASER using either estimated pedotransfer functions or measured hydraulic parameters. *Water Resour. Manag.* **2007**, *21*, 1567–1584.

9. Wang, T.; Zlotnik, V.A.; Šimunek, J.; Schaap, M.G. Using pedotransfer functions in vadose zone models for estimating groundwater recharge in semiarid regions. *Water Resour. Res.* **2009**, *45*, doi:10.1029/2008WR006903.

10. Van Genuchten, M.T. A closed-form equation for predicting the hydraulic conductivity of unsaturated soils. *Soil Sci. Soc. Am. J.* **1980**, *44*, 892–898.

11. Wang, T.; Franz, T.E.; Zlotnik, V.A. Controls of soil hydraulic characteristics on modeling groundwater recharge under different climatic conditions. *J. Hydrol.* **2015**, *521*, doi:org/10.1016/j.jhydrol.2014.12.040.

12. Okamoto, K.; Sakai, K.; Cho, H.; Nakamura, S.; Nakandakari, T. Influences of a bulk density for hydraulic conductivity and water retention curve of Shimajiri maji soil, and examination of PTFs' applicability. *J. Rainwater Catchment Syst.* **2015**, *21*, 1–6. (In Japanese with English Abstract)

13. Miyamura, N.; Iha, S.; Gima, Y.; Toyota, K. Factors limiting organic matter decomposition and the nutrient supply of soils on the Daito Ilands. *Soil Microorg.* **2011**, *65*, 119–124.

14. Classification of cultivated soils in Japan. Available online: http://soilgc.job.affrc.go.jp/Document/Classification.pdf (accessed on 4 September 2015).

15. Schaap, M.G.; Leiji, F.J.; van Genuchten, M.T. Neural network analysis for hierarchical prediction of soil water retention and saturated hydraulic conductivity. *Soil Sci. Soc. Am. J.* **1998**, *62*, 847–855.

16. Schaap, M.G.; Leiji, F.J. Improved prediction of unsaturated hydraulic conductivity with the Mualem-van Genuchten model. *Soil Sci. Soc. Am. J.* **2000**, *64*, 843–851.

17. Schaap, M.G.; Leiji, F.J.; Van Genuchten, M.T. Rosetta: A computer program for estimating soil hydraulic parameters with hierarchical pedotransfer function. *J. Hydrol.* **2001**, *251*, 163–176.

18. Kosugi, Y.; Katsuyama, M. Evapotranspiration over a Japanese cypress forest. II. Comparison of the eddy covariance and water budget methods. *J. Hydrol.* **2007**, *334*, 305–311.

19. Jayasinghe, G.Y.; Tokashiki, Y.; Kitou, M.; Kinjyo, K. Effect of synthetic soil aggregates as a soil ameliorant to enhance properties of problematic gray ("Jahgaru") soils in Okinawa, Japan. *Commun. Soil Sci. Plant Anal.* **2010**, *41*, 649–664.

20. Mualem, Y. A new model predicting the hydraulic conductivity of unsaturated porous media. *Water Resour. Res.* **1976**, *12*, 513–522.

21. Van Genuchten M.T. *A Numerical Model for Water and Solute Movement in and Below the Root Zone*; U.S. Salinity Laboratory, USDA, ARS, Riverside: CA, USA, 1987.

22. Van Genuchten, M.T.; Gupta, S.K. A reassessment of the crop tolerance response function. *Indian Soc. Soil. Sci.* **1993**, *41*, 730–737.

23. Smith, D.M.; Inman-Bamber, N.G.; Thorburn, P.J. Growth and function of the sugarcane root system. *Field Crops Res.* **2005**, *92*, 169–183.

24. Larsbo, M.; Jarvis, N. *MACRO 5.0: A Model of Water Flow and Solute Transport in Macroporous Soil*; Department of Soil Sciences, Swedish University of Agricultural Sciences: Uppsala, Sweden, 2003.

25. Nakama, M.; Nose, A.; Miyazato, K.; Murayama, S. *The Science Bulletin of the Faculty of Agriculture, University of the Ryukyus*, 1st ed.; Faculty of Agriculture, University of the Ryukyus: Naha-shi, Japan, 1987; pp. 187–198.

26. Penman, H.L. Natural evaporation from open water, bare soil and grass. *Proc. R. Soc. Lond. A* **1948**, *193*, 120–145.

27. Miura, T.; Okuno, R. Detailed description of calculation of potential evapotranspiration using the Penman equation. *Trans. Jpn. Soc. Irrig. Drain Reclam Eng.* **1993**, *164*, 163.

28. Campbell, G.S. *Soil Physics with Basics, Transport Models for Soil-Plant Systems*, 1st ed.; Elsever: New York, UN, USA, 1985; p. 144.

29. Schaap, M.G.; Leiji, F.J. Database-related accuracy and uncertainty of pedotransfer functions. *Soil Sci.* **1998**, *163*, 765–779.

30. Yanagawa, A.; Fujimaki, H. Tolerance of canola to drought and salinity stresses in terms of root water uptake model parameters. *J. Hydrol. Hydromech.* **2013**, *61*, 73–80.

8

Assessing Climate Change Impacts on Water Resources and Colorado Agriculture using an Equilibrium Displacement Mathematical Programming Model

Eihab Fathelrahman [1,]*, **Amalia Davies** [2], **Stephen Davies** [2] **and James Pritchett** [3]

[1] Department of Agribusiness and Consumer Sciences, College of Food and Agriculture, United Arab Emirates University, Al-Magam Campus, P.O. Box 15551, Al-Ain, United Arab Emirates

[2] Pakistan Strategy Support Program (PSSP), the International Food Policy Research Institute (IFPRI), IFPRI-PSSP Office #006, Islamabad, Pakistan; E-Mails: sdav45@hotmail.com (A.D.); stephen.davies@colostate.edu (S.D.)

[3] Department of Agricultural and Resource Economics, Colorado State University, Campus Mail 1172, Fort Collins, CO 80523, USA; E-Mail: James.Pritchett@ColoState.edu

* Author to whom correspondence should be addressed; E-Mail: eihab.fathelrahman@uaeu.ac.ae;

Abstract: This research models selected impacts of climate change on Colorado agriculture several decades in the future, using an Economic Displacement Mathematical Programming model. The agricultural economy in Colorado is dominated by livestock, which accounts for 67% of total receipts. Crops, including feed grains and forages, account for the remainder. Most agriculture is based on irrigated production, which depends on both groundwater, especially from the Ogallala aquifer, and surface water that comes from runoff derived from snowpack in the Rocky Mountains. The analysis is composed of a Base simulation, designed to represent selected features of the agricultural economy several decades in the future, and then three alternative climatic scenarios are run. The Base starts with a reduction in agricultural water by 10.3% from increased municipal and industrial water demand, and assumes a 75% increase in corn extracted-ethanol production. From this, the first simulation (S1) reduces agricultural water availability by a further 14.0%, for a combined decrease of 24.3%, due to climatic factors and related groundwater depletion. The second simulation (S2-WET) describes wet year conditions, which negatively affect yields of irrigated corn and milking cows, but improves yields for important crops such as non-irrigated wheat and forages. In contrast, the third simulation (S3-DRY) describes a drought year, which leads to reduced dairy output and reduced corn and wheat. Consumer and producer surplus losses are approximately $10 million in this simulation. The simulation results also demonstrate the importance of the modeling trade when studying climate change in a small open economy, and of linking crop and livestock activities to quantify overall sector effects. This model has not taken into account farmers' adaptation strategies, which would reduce the climate impact on yields, nor has it reflected climate-induced shifts in planting decisions and production practices that have environmental impacts or higher costs. It also focuses on a comparative statics approach to the analysis in order to identify several key effects of changes in water availability and yields, without having a large number of perhaps confounding assumptions.

Keywords: water demand; Colorado Equilibrium Displacement Positive Mathematical Programming model (Colorado EDMP); small open economy; food and energy

---

1. Introduction

The agricultural economy in Colorado is dominated by livestock production and sales, which account for 67% of total receipts. Crops, including feed grains and forages, account for the remainder. Most cropping receipts are based on irrigated production, which is sourced from groundwater, especially from the Ogallala aquifer, and surface water, which comes from runoff of snowpack in the Rocky Mountains. Currently, about 86% of water resources are in agriculture, but this is projected to decline due to demographic factors that lead to increased Municipal and Industrial (M&I) water demand, economic factors related to higher costs of irrigation, increased water demand for oil shale mining, and geographic factors such as climatic changes and groundwater depletion. Moreover, hydrologic studies point to an expected decline in runoff from 6% to 20% by 2050, and also a shift in the timing of that runoff to earlier in the spring. These studies also showed that late-summer flows may be reduced [1–4].

Colorado agriculture has blossomed with the development of water resources used for growing crops, which, in turn, spurs value-added production in the meat and dairy subsectors. Yet, increasing urban development is expected to create a reallocation of 740 million m^3 (hereafter million = M) of agricultural water to new municipal and industrial demands by 2030 [5]. Another challenge to the agricultural sector is a possible expansion of ethanol production in Colorado. Shifting corn to ethanol use rather than animal feed could place livestock production, Colorado's dominant agriculture industry, at a disadvantage as the key input becomes more expensive, even though dry distillers' grains mitigate some of the constraint. These pressures on agriculture may be exacerbated by the presence of climate change, particularly its effect on water availability and yields. Stakeholders thus seek ways to better understand the implications of climate change on statewide water availability and requirements for crops and livestock, in the presence of a larger population and other new demands such as ethanol production. This research evaluates these issues with illustrations on how resources might be reallocated and how prices respond in the future.

The research uses a positive mathematical programming model specified to represent the Colorado agricultural sector, which is simulated to examine impacts of selected future constraints on water and yields resulting from climate change. First, this model was calibrated to 2007 quantities and prices. Then, a "Base" scenario was constructed, which reflects several future drivers of change affecting the state's agriculture: (1) increasing competition for water due to population growth, especially shifts in the resource from agricultural to municipal uses in the South Platte and Arkansas River basins; and (2) we also add two ethanol plants into the South Platte River Basin, which leads to a 75% increase in corn extracted-ethanol production there, and provides competition to the cattle feeding industry's use of a key input, corn for grain.

The changes incorporated into the Base scenario are related to the anticipated growth in the local economy, but to do not include effects of climate change. With this Base established, we run three simulations to explore the implications of climate change. The first one further reduces water availability based on forecasts of reduced runoff, while the second and third simulations introduce yield changes that might arise due to higher temperatures and increased variability of rainfall. Results for these scenarios are reported in terms of acreage changes, total value of production, exports and imports from the state, and prices. The overall changes in consumer and producer surpluses across the simulations are also reported. This modeling effort does not attempt to capture the full set of dynamic effects that will in fact occur, because for a small region, the range of possible outcomes over the next several decades is high, and is dependent on an equally extensive set of possibilities. Our approach is thus to focus on important outcomes with regard to climate change using a comparative statics method.

The document is organized into a series of sections. The current status of Colorado agriculture and its dependence on irrigation water supplies is reviewed in Section 2. This section also includes a review of expected climate change impacts on the availability of water and effects on commodities. Section 3 provides a literature review with regard mathematical programming methodology, while Section 4 lays out our particular model. Section 5 provides a discussion of the simulations and results, and Section 6 gives conclusions and thoughts for further research.

2. Colorado Agriculture and Water Use: Current and Projected Changes

This section contains two parts: the first covers the current size and structure of Colorado agriculture and describes key changes that might occur over the next decades; the second looks at the current pattern of water use and reviews forecasts of water reallocation.

2.1. Agriculture in Colorado

The agricultural economy in Colorado is dominated by livestock (almost $5.8 billion in sales during 2007, the year used to calibrate our model), which accounts for 67% of total receipts from the sector. The 2007 commodity balances are contained in Table 1. Colorado agriculture is heavily traded outside the state and abroad, as we learned when building commodity balance sheets used in the model. Fed beef, the largest economic sector, produced $3.4 billion in 2007 and traded 82% of its production out of state. The cattle feeding industry creates a substantial derived demand for corn production ($463 Million hereafter M) and corn imports, which reached $703 M in the same year. In

2007, 75% of the total value of Colorado's crops came from irrigated acreage, as most of hay, corn, and pasture for livestock were produced on irrigated land [6].

Table 1. Production, in-state sales, exports, and imports of key Colorado agricultural commodities.

Crop or Commodity	Production (M $)—Column 1	In State Sales (M $)—Column 2	Exports (M $)— Column 3	Imports (M $)— Column 4	% Exports/ production % of column 3 /column 1	% Imports/ production % of column 4 /column 1
Corn*	462.8	1051.7	113.7	702.6	24.6	151.8
Wheat	483.5	61.0	474.8	52.3	98.2	10.8
Barley	185.3	67.2	163.9	45.9	88.5	24.7
Sorghum	383.7	400.3	0.0	16.6	0.0	4.3
Dry beans	24.7	7.2	17.6	0.0	71.0	0.0
Beef	3382.5	905.6	2748.4	0.0	81.3	0.0
Cow calf	135.8	278.2	0.0	142.4	0.0	1.0
Hogs	170.9	204.0	0.0	33.1	0.0	19.3
Dairy	522.3	566.4	0.0	44.1	0.0	8.5
Sheep	488.5	27.1	461.4	0.0	94.5	0.0
Broilers	145.7	205.4	0.0	59.6	0.0	40.9
Eggs	74.1	85.6	0.0	11.5	0.0	15.5

Note: * Corn sales includes ethanol production.

It is not possible to say how much imported corn went into ethanol production, but ethanol used the equivalent of 23% of the state's production, while 67% of corn was imported. The value of wheat production equaled that of corn output, but 98% was exported across state boundaries. The sheep and lamb industry is also heavily export-oriented, with slightly less than $500 M in revenues during 2007, and 94% exported. Sorghum was the largest feed grain produced after corn, with revenues in excess of $380 M and imports totaling about $17 M. Instate sales of corn, excluding the ethanol industry, exceeded $800 M, while sorghum was $400 M. Colorado's dairy and hog sectors sold output within the state and required imports to meet demand, totaling 8.4% and 19% of production respectively. Imports of cows and calves were 50% of instate calf sales ($278 M) with buyers almost exclusively being feedlots. Barley and dry beans were relatively small agricultural subsectors and produced mostly for exports (88% and 71% of their production respectively). At the other end, 30% of broilers' sales in Colorado (about $60 M) and 13% of egg sales were imports.

Ethanol production may play a key role in Colorado's energy future and plans therefore exist to expand production capacity. Yet, Colorado is a small producer of ethanol, with just three plants located in the South Platte River Basin. The average plant capacity in Colorado is 215 M liters per year, or about 1.3% of the nation's ethanol capacity. The "corn footprint," or demand by these plants, is approximately 1.6 M tons each year, which requires about 130,000 hectares of irrigated corn production.

Expected Climate Change Effects on Colorado Agriculture. A consensus of climate change models suggest temperature in Colorado is expected to increase by up to 9–11 degrees Fahrenheit in the worst case scenario. The timing of seasons is likely to shift as well, with an earlier spring and longer fall. Midwinter precipitation should occur later in the calendar year, while less rain is expected to fall in late-spring and summer. As temperatures rise, runoff will peak earlier in the spring and be reduced significantly in late summer. Earlier run off could result in an 8.5% reduction of in-stream flows by midcentury in the Colorado River basin and a 5%–10% possible reduction in the Arkansas and Rio Grande basins. Little work has been done for the South Platte in terms of the impact of climate change on winter snow runoff [7]. The variability year to year is also likely to grow.

Climate change will have effects on crop yield and water requirements. The main climate factors affecting agriculture are temperature, availability of water, and the concentration of atmospheric CO_2. Soil water availability depends on the above three factors as they interact with soil properties, while field humidity, clouds and solar radiation also influence plant water requirements. The major commodities in Colorado agriculture are affected variously by these climate factors. For corn, the yield loss associated with increased temperature exceeds the positive effects of increasing carbon dioxide levels, so yields are expected to decline [8]. Also, high temperatures earlier in the season lead to less pollen germination and lower yields [9]. The changing precipitation patterns suggest increased yields for non-irrigated wheat in Colorado given the increase rainfall in winter and early spring.

High temperatures also extend the number of growing degree days in the crop season, which has a positive effect on yields and overall production for hay. However, few studies exist on the effects of climate change for this crop. In a review of three studies, depending on the assumed increase of CO_2 concentration, alfalfa yields were estimated to change from a 16.7% increase to a decrease of 19.4%. However, this added growth and length of season may lead to lower nutritional content, depending on soil quality constraints [10,11]. On the other hand, productivity may be higher than previously expected in semi-arid grasslands, and thus additional forage may become available [12].

Warmer temperatures increase plant evapotranspiration, while CO_2 concentration partially offsets this process by increasing plant water-use efficiency. Wheat and hay are more sensitive to CO_2 than corn [8,13]). Although there is great uncertainty about the future CO_2 concentration, it is unlikely to neutralize the effect of anticipated, protracted droughts on crop production.

Increasing heat also affects livestock growth and performance. Higher temperatures reduce livestock production in the summer but increase it in winter. Under heat stress, animals reduce grazing to stay in the shade, thus reducing their feed intake and suffering from weight loss. Reduced quality of forage and digestibility leads to reduced dairy productivity. The greater the stress, the easier is the spread of parasites and disease pathogens. For dairy cows, heat stress reduces the milk fat and protein content in milk, and the quantity of milk produced is reduced up to 10%; moreover, other factors may also lead to lower yields as high-producing dairy cows are the most susceptible to heat stress due to breeding selection for high productivity, and reproduction rates are also adversely impacted [3,14–16].

2.2. Colorado's Outlook for Water Resources

Competition for water is increasing in the West. Colorado is a headwater state, supplying water through river systems to eighteen downstream states. Interstate compacts mean that Colorado is not entitled to all surface water flows, and may only retain six billion m^3 in an average year. This water is allocated among users according to the Prior Appropriation Doctrine, and, as nearly all of Colorado's rights have been appropriated, new users must obtain rights from others through voluntary transactions. Agriculture is the largest diverter and consumptive user of these surface flows. Agriculture also makes use of groundwater resources so that, on average, 1.0 M hectares of cropland are irrigated via groundwater or surface water. As noted earlier, irrigated crops comprise three-quarters of cropping receipts in Colorado, with two-thirds of these receipts bound for Colorado's livestock feeding industry [17].

Irrigation water depends on both groundwater, especially from the Ogallala aquifer, and surface water, which comes from runoff due to snowpack in the Rocky Mountains. Currently, about 86% of the state's water resource is used in agriculture, but this amount is projected to decrease. Causes for decline include demographic factors, such as increased Municipal and Industrial (M&I) water demand, economic factors related to higher costs of irrigation, increased water demand for oil shale mining, and geographic factors such as climatic changes and groundwater depletion.

While agriculture holds the majority of water rights, new demands for water resources come from a growing population and environmental uses. Population forecasts are for an increase of more than 50% in the next twenty years, so a gap between existing municipal water supplies and demand from the larger population is anticipated. The Colorado Water Conservation Board's Statewide Water Supply Initiative (SWSI) predicts that Colorado's South Platte Basin will experience a 61.9% increase in water demand, or about 505 M m^3, by 2030, which will continue to rise thereafter. With water already appropriated in the South Platte, an estimated 73,000 irrigated hectares will need to be permanently fallowed to supply these increasing demands. The plans for nearly all South Platte water providers include significant water rights transfers [1,18].

Great variation exists among findings of hydrologic studies regarding expected decline in runoff, from 6% to 20% by 2050, although there is consensus on the persistence of the shift of runoff to earlier in the spring, and a change in precipitation to a greater intensity during winter and lesser in spring and summer [1–3]. The topography of the state and other factors make projections particularly complex [19].

3. Literature Review

The model used in this research is an optimization model using mathematical programming in a manner that has a long history in economics and engineering. The approach chooses activity levels that maximize an objective function in the face of physical constraints on resources. Positive Mathematical Programming (PMP) improves on earlier techniques by allowing perfect calibration to a base and additions of more realistic behavior into such models [20,21]. As an activity based approach, PMP simplifies communication across disciplines and is particularly suited to study bio-physical and environmental features of agricultural systems.

Over the last 10 years, the PMP approach has been object of extensive review, critique and extensions [22–25], as policy makers increased their reliance on quantitative economic models to understand effects of agricultural policies. As such, the method has been widely used in sectoral and regional analysis. In the European Union (EU), several models analyzed policy instruments within the EU's Common Agricultural Policy (CAP), especially the effects of the CAP reform starting in 2003–2004, where a switch to decoupled payments to farmers was made. Some examples of these models include the FAL, Parma and Madrid models, which use PMP to calibrate to observed values, and also apply the maximum entropy approach to estimate total variable costs [26–36].

The PMP method is thus versatile enough to model policy scenarios in a straightforward fashion, and has been adopted as especially well-suited to examine animal feed requirements and land constraints [25], and to study jointly agricultural outputs and environmental externalities [31]. Howitt *et al.* [32] applied the methodology to estimate effects of climate change on irrigated agriculture in California using the State Water and Agricultural Production model (SWAP). SWAP improves on traditional PMP models by allowing for large policy shocks and enhanced flexibility in handling input substitutions. These models are often linked to hydrological network models and other biophysical system models.

The equilibrium displacement modeling approach [33,34] represents an economic system of demand and supply relationships, and can show the effects of exogenously determined shifts of supply and demand from an initial equilibrium (a displacement). Changes in market prices and quantities resulting from the displacement determine changes in consumer and producer surpluses. This follows originally from Samuelson [35], who shows that maximizing profits is equivalent to maximizing the total surplus when markets are competitive.

The Equilibrium Displacement Mathematical Programming (EDMP) model originally developed by the USDA Economic Research Service Harrington and Dubman [36] is a sector-wide, comparative statics model of the U.S. agricultural sector, applying a mathematical programming approach to the equilibrium displacement methodology, with specific farm sector relationships and policies reflected. They used values estimated by econometric studies and applied the asset-fixity theory of Johnson and Quance [37] to estimate slopes of supply functions. The Harrington and Dubman model is similar to the general PMP approach, but the supply and demand curves are explicit, and the base calibration is achieved by shifting intercepts until they match initial values with as much precision as is needed. Thus this approach is termed an "equilibrium displacement mathematical programming" model.

Regional and Climate Change Studies. Connor *et al.* [38] noted that an increasing number of analyses assess the impacts of climate change on irrigated agriculture in arid and semi-arid regions of the world, especially those that face a projection of drier weather. The objective function of their irrigation sector model maximizes profits across three sub-regions in the Murray-Darling River basin, Australia, subject to land and water constraints. The scenarios included a base case, a water scarcity model, a water variability model, and full effects model. The latter model includes both water variability and implications for changes in salinity. They concluded that ignoring the combined water-climate effects, along with salinity, leads to results that understate costs and impacts on output. Moreover, using the analysis of salinity, they identify various thresholds of climate change that create structural change in productivity and costs related to levels of salinity.

Henseler *et al.* [39] studied global change in the Upper Danube basin using an agro-economic production model, with two climate change scenarios. The first scenario assumed a significant increase in temperature, while the second one showed effects of a moderate increase. This study's results showed large differences in agricultural income and land use between the two scenarios and shifts that lead to increases in cereal production and extensive grassland farming due to the increased temperature in the first scenario. Qureshi *et al.* [40], Whitney and van Kooten [41], and Wolfram *et al.* [42], studied climate change impacts on agriculture at the regional levels in Canberra Australia, Western Canada, and California respectively. These studies reached conclusions that are similar to the studies discussed above. Whitney and van Kooten [41] expanded the model to include impacts on pasture and wet-land.

Finally, with regard to previous Colorado analyses, Bauman *et al.* [43] estimated the economic impacts of the drought in 2011 using an Input–Output (I/O) model and a variant of the current

Colorado Equilibrium Displacement Model. The authors found that the 2011 Colorado accounted for $83 to $100 M in economic impact, when all economic sectors of the state economy were included. Schaible *et al.* [44] argued that the gradual warming in the Western United States is expected to shift the precipitation pattern and alter the quantity and timing of associated stream flows. In addition, the effects of climate change will move bio-energy growth to the Ogallala aquifer in the Western States, which demand that careful optimization of water use is needed to choose irrigation technologies. They underline the importance of further research to understand economic implications of climate change at the regional level.

Thus, previous studies agreed that there are likely to be significant shifts in land use and crop mix due to climate changes at the regional level. These studies also agreed on the importance of understanding possible structural changes, and noted that there will be significant income and price effects due to climate change. Furthermore, the above review suggests that a lack of studies investigating the impact of climate change at the regional level exist, in particular those that trace out impacts in a small, open economy via trade with the Rest of the World (ROW) and include livestock and crop interactions. Previous studies also agree that positive mathematical modeling fits the research problem and unveils opportunity to simulate possible production and cost changes due to climate change, which should enable a better understanding of welfare implications at the regional level.

4. Structure of the Colorado Equilibrium Displacement Positive Mathematical Programming (Colorado EDMP) Model

The Colorado Equilibrium Displacement Positive Mathematical Programming model (Colorado EDMP) is a variant of the EDMP model by Harrington and Dubman, which the authors adapted for Colorado's agricultural sector [45]. This model maximizes the sum of producer and consumer surpluses across most major products in Colorado's agricultural sector, subject to a number of spatial market and resource constraints. The Colorado EDMP is calibrated to Colorado's agricultural economy, and adds other natural resource dimensions (*i.e.*, Colorado agricultural sector demand for water). Spatial constraints consist of three regions with separate water availability for irrigation in each basin (South Platte River basin, Arkansas Basin, and other Colorado basins) along with differing crop water requirements in each basin. These requirements were developed using irrigation water requirement (IWR) coefficients per crop per region from the Colorado Decision Support System (CDSS) weather and soil characteristics databases [46]. The optimization model selects food and feed crops, water supplies, and other inputs to maximize the sum of producer and consumer surpluses, subject to constraints on water and land, and subject to economic conditions regarding prices, yields, and variable costs. In the following paragraphs, we describe the Colorado EDMP and its basic dimensions.

The particular function given below is a second order Taylor series expansion as first introduced by Takayama and Judge [47], which permits an approximation of an unknown functional form for the cost function:

$$\text{Max: } Z = F'x - 1/2\, x'H\,x \tag{1}$$

with $x > 0$, where x is a vector of endogenous variables that relate to sector demand and production processes. In the following expanded form of the Equation (1), the vectors x are divided into five groups. In the notation below the vectors of variables are written in lower case, while the vectors of parameters are in upper case, and indices under the summation operators are simplified as:

$$Z = \sum_j (F'' - .5H\, q_j)q_j - \sum_i \sum_b (F'' + .5Hcl_i)cl_i - \sum_n u + \sum_g (F'' - .5He_g)e_g - \sum_s (F'' - .5HM_s)M_s \qquad (2)$$

where, q_j = domestic sales of j agricultural commodities (in M tons) and livestock products (M head, tons, or dozens of eggs); cl_i = feed and food crop activities i identified by river basin (for selected crop activities, in M hectares) and livestock activities (head counts, live weight, milk tons and dozens); u = dollar value of n inputs (in M dollars); e_g = exports of t agricultural commodities (in M tons) and livestock products (M tons, dozens of eggs); M_s = imports of s agricultural commodities (in M tons) and livestock products (M tons, dozens of eggs); F'' = a vector of intercepts indexed under each set above, which are determined in the calibration phase; H = the diagonal elements of the Hessian matrix flowing from the First Order Conditions. H is assumed to be negative semi-definite.

In Equation (2), the first term is the function of total revenue, where $(F'' - .5H\, q_j)$ = p is the vector of price dependent domestic demand functions, and p is the vector of output prices. The H_j elements are derived from predetermined elasticities of demand for j commodities and livestock products. The second element is a non-linear total variable cost function, where H_{ib} are elements of the Hessian of supply functions; they are calculated as the ratios of capital replacement costs over excess capacity for i activities in b river basins. The term $(F'' + .5Hcl_i)$ = Marginal Cost provides the supply side equivalent to a price dependent demand function in the first term. The third element is the sector's sum of inputs used in the sector, entered in value terms. The last two elements represent the export and import functions (these include out-of-state trade as well as international trade), which are included in the sector's the objective function (see also Helming [48]). Ht and Hs are also exogenously calculated. Examples of the constraints included in the mathematical program are presented in Appendix.

The agricultural activities in the model cover 91% of total agricultural production in Colorado, including thirteen crop and nine livestock commodities, which are sold to local consumers or out-of-state exports. Imports for nine products are present and compete with local production. The nine livestock sectors are cow calf, fed beef, hogs, dairy, sheep, broilers and layers, turkeys, and horses. Some of these livestock activities produce multiple products, including meat, milk, and/or eggs. Demand for feed crops and forages are derived from livestock activities through demand for rations. Food crops are wheat, potatoes, sunflower, and dry beans. Calf imports go directly into the cattle feeding industry. The commodities included, their acreage and production values, and a comparison of how our calibrated model compares to historical 2007 values is given in Appendix Table A1.

The model also includes accounting costs for all activities. Inputs are categorized in the following categories: genetic inputs, such as seed or calves; specialized technology; mineral fertilizers (without manure applications); other chemicals; fuel and lube; electricity; irrigation energy and other irrigation costs; other variable purchased inputs; fixed cash costs; and capital replacement costs. Farm production costs reflect various yields and cost structures in different basins. Irrigated and non-irrigated crop costs are derived from enterprise budgets created by extension professionals in Colorado and the High Plains. Currently, the relationship between inputs and outputs is fixed, with no substitution, so that corn production, for example, has a fixed yield of 8.3 tons per hectare and each hectare uses a certain quantity of fertilizer, other chemicals, and irrigation energy (when irrigated).

Demand elasticities from the literature provide the values for the Hessian's elements related to demand, which help the model, provide reasonable responses when used in scenario analyses. The F values, or intercept terms, are estimated by repeated adjustments until the prices and quantities are calibrated to a desired level of accuracy.

It is possible, with enough time, to exactly calibrate prices and quantities by shifting demand and supply intercepts. While this can be a tedious process, it provides an examination of the relationships and tendencies in the model, which cannot be achieved as intuitively when using a large set of cross price elasticities that, in any case, cannot be reliably identified for a small region like Colorado. We show the results of our efforts at calibration in Appendix Table A1, where the table shows the calibrated quantities *versus* actual values for selected products. It also provides estimates of the intercepts and slopes (Hessian elements) of the associated supply curves.

5. Base Scenario and Climate Change Simulations

This research includes three climate change simulations that are compared to the Base simulation, where the "Base" is designed to represent selected features of the sector several decades in the future. The two main features included are reduced water availability in the South Platte and Arkansas River basins, and added demand for ethanol, which represents a competing demand for corn. Three simulations then are created to show incremental effects of climate change on the Base model. The first simulation (hereafter S1) reduces agricultural water availability by a further 14.0% across all basins, for a combined decrease of 24.3%. This reduction comes from climatic factors and related groundwater depletion, as detailed in the Colorado Water Conservation Board's Statewide Water Supply Initiative study (CWCB) [2]. There are no changes in yields or other factors.

In addition to the direct water reduction, the effects of increased heat and an extreme dry year are reflected in the second and third simulations. First, climate change models suggest up to a 9–11 degree Fahrenheit increase in temperature, as a high end case [23]. The average rise in temperature also affects the variability and likelihood of years with more extreme weather, as illustrated in Figure 1. This figure illustrates how the increase in average temperature leads to a greater likelihood of extreme weather events, such as droughts, but also to years with higher precipitation. Simulation two represents a warm and wet year (hereafter S2-WET), with shifts in the pattern of precipitation, but with an increase in average temperature included as well. The third simulation reflects a drought year (S3-DRY) with dry conditions, in addition to the temperature increase and shifts in precipitation found in S2-WET.

Table 2 summarizes the percentage changes in crop yields and dairy productivity from those used in S1. The irrigated corn yield in S3-DRY decreases due to higher July temperatures, and from lack of rain and cloud cover, which hampers pollination [8]. Yields for non-irrigated wheat increase as sufficient winter rainfall is present s during the critical growing period in S2-WET, and decline by an equal amount in S3-DRY to reflect the effect of less rainfall and higher temperatures [13].

Both irrigated hay and corn silage yields surge with higher temperatures, which result in a longer growing season and more cuttings, in the case of hay, and help biomass growth in silage. Because both are grown on irrigated land, decreased rainfall does not have an effect, and yields are kept high in both scenarios. Yields in rangeland and pasture increase in S2-WET year, as sufficient rainfall supports germination and growth, but like other non-irrigated crops, these sources of feed see reduced yields in

S3-DRY. Dairy sector productivity plummets in both the second and third simulations, reflecting animal stress from high temperatures in absence of mitigating strategies. (These impacts of climate change are presented in more detail in Section 2).

Table 2. Percent yield and productivity changes in S2-WET and S3-DRY, relative to S1. Sources: [13] (pp. 34–48, 56–61, 77–82).

Simulation	Irrigated corn	Dryland wheat	Irrigated hay	Silage	Pasture	Rangelands	Dairy
S2-WET	−0%	13%	18%	13%	8%	8%	−18%
S3-DRY	−15%	−13%	18%	13%	−13%	−13%	−18%

In summary, the following conditions are analyzed in the next sections:

Base: This scenario examines the economic impacts of shifting water resources from agricultural to municipal uses in the South Platte and Arkansas River basins by 22% and 18% respectively;

S1: This simulation alters the Base scenario by reducing agricultural water availability by a further 14.0% across all basins for a combined decrease of 24.3% based on expected climate change effects;

S2-WET: This simulation represents a warm and wet year, with shifts in the pattern of precipitation and an increase in average temperature;

S3-DRY: A drought year is simulated in the third example, using dry conditions along with the temperature increase and shifts in precipitation found in S2-WET.

Figure 1. Climate change scenarios in Colorado Economic Displacement Mathematical Programming (EDMP).

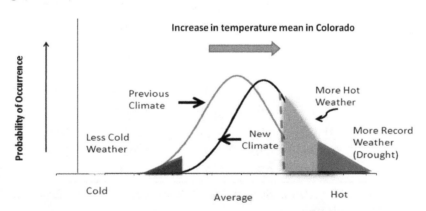

Note: Relatively small shift in the average climate can substantially increase risk of extreme events such as drought [4].

Base Scenario. The Base scenario reflects selected supply and demand factors for agricultural inputs and outputs in the future. First, it includes expected implications of competition between the agricultural sector and other sectors (e.g., M&I) for water at the basin level. In particular, this scenario shifts water resources from agricultural to municipal uses in the South Platte and Arkansas River basins by reducing water availability to agriculture by 22% and 18% respectively, with respect to calibrated values for 2007. This reduction follows estimates by the Colorado River Water Availability Study—CRWAS-report [49], and results in the fallowing of a proportional amount of irrigated land in each basin, although individual crops can vary without constraint aside from the overall reduction in irrigated acreage. Overall a net decrease of 10.3% in total water availability occurs because nearly 50% of annual volume is in river systems outside of these two basins.

Additionally, this simulation adds two ethanol plants in the South Platte River Basin, thereby increasing Colorado ethanol production from 175 M gallons annually to 308 M gallons. The Base scenario values are found in Tables 3 and 4.

Base Scenario Results. Due to an anticipated reallocation of water from agriculture to municipal uses in the South Platte and Arkansas basins in the Base scenario, crop acreage shifts relative to the calibrated values of 2007, particularly for those commodities that are produced on irrigated land. Also, adding two ethanol plants raises annual production from 662 M liters annually to 1165 M liters. This increase raises demand for corn by about 1168 k tons (hereafter thousand = k), which must be supplied from various sources. On the one hand, other uses of corn can decrease, which in this model are feed, final consumption and exports. Also, supplies can come from added production or greater imports.

Table 3. Area harvested in Base scenario and Climate Change Simulations.

Crop/Livestock product	Base	Simulation			Percentage change from Base		
		S1	S2-WET	S3-DRY	S1	S2-WET	S3-DRY
South platte dry corn	0.07	0.07	0.06	0.1	0.0%	−14.3%	42.9%
South platte irrigated corn	0.27	0.26	0.25	0.17	−3.7%	−7.4%	−37.0%
Arkansas irrigated corn	0.07	0.07	0.07	0.04	0.0%	0.0%	−42.9%
Arkansas dry corn	0.16	0.17	0.15	0.04	6.3%	−6.3%	−75.0%
All corn	0.58	0.57	0.53	0.36	−1.7%	−8.6%	−37.9%
South platte dry wheat	0.62	0.65	0.65	0.33	4.8%	4.8%	−46.8%
South platte irrigated wheat	0.03	0.03	0.04	0.16	0.0%	33.3%	433.3%
Arkansas dry wheat	0.29	0.29	0.3	0	0.0%	3.4%	−100.0%
All wheat	0.94	0.96	0.99	0.49	2.1%	5.3%	−47.9%
Other crops	0.17	0.17	0.17	0.17	0.0%	0.0%	0.0%
Colorado basin hay	0.35	0.28	0.28	0.28	−20.0%	−20.0%	−20.0%
South platte dry hay	0	0	0	0.08	0.0%	0.0%	0.0%
South platte irrigated hay	0.03	0	0	0	−100.0%	−100.0%	−100.0%
Arkansas irrigated hay	0.04	0.03	0.03	0.05	−25.0%	−25.0%	25.0%
All hay	0.41	0.31	0.31	0.41	−24.4%	−24.4%	0.0%

Notes: All values are in M hectares. Source: Model Runs from Colorado Economic Displacement Mathematical Programming (CEDMP) Model.

Sales of the main user of feed, fed beef, do not change much from the calibration to the base, even though water supplies have dropped by 10.3%. Significant and numerous changes in the feed sources, however, do occur. Corn sales to local users other than for livestock feeding decline by about 101 k tons from 2007, while a 3% reduction occurs in corn used for feed, or nearly 177 k tons arises, mainly in a shift to other, smaller grains that use less water. Exports decline by about 25 k tons as well. These shifts together release corn from other uses for a quarter of the increased ethanol demand. However, imports decrease by about 355 k tons, so overall supply is lower from these shifts and cannot fully support growth in corn demand for ethanol, as the variation in exports and imports just offset each other. Thus, production growth is the main source of supply for the increased demand for corn.

Table 4. Production, sales, exports, imports, and prices in the base and simulation scenarios.

Commodity	Variable	Base	Simulation			Percentage change from Base		
			S1	S2-WET	S3-DRY	S1	S2-WET	S3-DRY
Corn	Production	4567	4468	4265	2428	−2.2%	−6.6%	−46.8%
	Sales in Colorado	84	84	84	79	0.0%	0.0%	−6.1%
	Exports	699	693	678	549	−0.7%	−2.9%	−21.5%
	Imports	4128	4194	4326	5532	1.6%	4.8%	34.0%
	Prices [a]	145	145	145	160	0.0%	0.0%	10.0%
Wheat	Production	2251	2294	2722	1402	1.9%	20.9%	−37.7%
	Sales in Colorado	269	269	278	253	0.0%	3.0%	−6.1%
	Exports	2243	2281	2675	1461	1.7%	19.3%	−34.8%
	Imports	261	259	231	313	−1.0%	−11.5%	19.8%
	Prices [a]	220	228	220	243	3.3%	0.0%	10.0%
Fed Beef	Production	1389	1384	1375	1285	−0.3%	−1.0%	−7.5%
	Sales in Colorado	340	340	340	336	0.0%	0.0%	−1.3%
	Exports	1049	1044	1035	949	−0.4%	−1.3%	−9.5%
	Prices [a]	2644	2724	2729	2773	3.0%	3.2%	4.9%
Dairy	Production	1244	1244	1030	1030	0.0%	−17.2%	−17.2%
	Sales in Colorado	1357	1357	1266	1262	0.0%	−6.7%	−7.0%
	Imports	113	113	232	232	0.0%	104.0%	104.0%
	Prices [a]	429	402	500	500	−6.2%	16.6%	16.6%
Hay	Production	3053	2255	2273	3038	−26.1%	−25.5%	−0.5%
	Sales in Colorado	3830	4967	4032	4312	29.7%	5.3%	12.6%
	Imports	777	2712	1759	1274	249.0%	126.4%	64.0%

Notes: [a] Units of Prices are in $/ton; All other values are in k tons. Source: Model Runs from Colorado Equilibrium Displacement Mathematical Programming (CEDMP) Model.

The growth in production is nearly 1041 k tons, which comes from an increase of close to 152 k hectares in corn. This increase is generally in irrigated land in the Arkansas and South Platte basins, but in the Arkansas basin, a significant proportion of the production growth comes on non-irrigated land. Given that irrigated land is withdrawn from production in the Base scenario, growth in corn production must come from a shift out of other crops. This includes a reduction of area harvested for alfalfa hay by nearly one third, or about 223 k hectares, and a reduction of fallow land in the Arkansas basin. This occurred even though hay area in Other Colorado outside the two basins under consideration remained at about 315 k hectares.

The reduction in hay acreage is logical, based on its high water demand, significant use of irrigated land, and the possibility of using imports as a substitute. The irrigated corn area harvested in the South Platte basin increases by about 17.4% (or about 40.5 k hectares) from the calibrated value of 234 k hectares. Another 40.5 k hectares of alfalfa hay, or about 75% of its total area, is lost in the South Platte basin in response to limited water availability. The decline in irrigated corn and hay production negatively influences fed beef operations because locally produced feeds become more expensive.

Several general comments about the Base scenario are worth noting. First, given that most changes in the Base assumptions affect irrigated land, little reallocation occurs in non-irrigated products, such as wheat. While some shifts are found in wheat location, in the aggregate, its area drops by just under 2.0%. Despite the drop in acreage and production, exports, the main use of wheat, rise by about 2.0% or 35 k tons. This is possible mainly because of a shift of local sales of wheat into exports (71 k tons), a small increase in imports (17 k tons), which together permit a growth in exports despite the reduced production.

A second point is that changes in sales, production and consumption of other crops and livestock products occur relative to the calibrated model representing 2007, but for the remainder of this paper, these are not considered in depth. Our focus will be on cattle feeding and dairy, and their inputs, primarily corn and hay, and on wheat as the major non-irrigated product. These commodities account for 75% of area harvested and in-state sales, and about 85% of exports in the 2007 calibration.

The above scenario is created only by withdrawals of water from agriculture due to greater municipal and industrial uses, along with the presence of a larger ethanol industry. This clearly leaves out many possible changes that will occur in the next several decades, with the main ones being technological change and greater population. To reflect these changes, which some models attempt to do, we would need to make assumptions of a wide range of yields and productivity of livestock and dairy, and the increase in consumption of all products from the larger population. This seems to us to be a relatively non-productive effort for a small region like Colorado. Thus our Base is a mixture of the 2007 setting, with selected future effects made to key variables. The proportions of imports and exports stay roughly the same, even though they are not fixed, because the balance between demand and supply is not forecasted into the future. While it is certainly not an exact representation, the Base case permits us to examine important effects of climate change on yields and water availability, without being confounded by added, perhaps unsupportable, changes. Thus, the following results show additional effects due to water and yield changes coming from climate change.

6. Climate Change Scenario Results

As described earlier, three climate change simulations are included in this study. The following discussion of results is split into two sections, where the first summarizes and explains shifts in area within each simulation, which are presented in Table 3. These area shifts are related to a series of price effects that lead to additional variation in feed use, production levels, and exports and imports. These added effects of climate change are found in Table 4.

Simulated Area Effects. Relative to the Base, the area harvested of Colorado *corn* (about 600 k hectares) only changes slightly in S1. In S2-WET, overall area harvested declines by nearly 8%, but the change is not distributed equally across basins. The largest change in cultivated area occurs in S3-DRY, as total harvested area drops by 38%. This decrease is similar across both regions and for irrigated land, as the percentage decline is nearly identical in both the South Platte and Arkansas basins. The greatest impact in S3-DRY occurs in the Arkansas basin's non-irrigated land (−75%), which drops to only 45 from 151 k hectares. Conversely, South Platte non-irrigated corn expands by 11.8% over S2-WET, responding to higher prices coming from the large reduction in irrigated corn area harvested. The drought-like conditions in S3-DRY with high heat cause a large reduction in irrigated corn harvested as yields decrease by 8% from the Base. These results indicate the high sensitivity of corn area to variations created by climate change.

The total area harvested of *wheat* in Colorado (with a baseline of 0.94 M hectares) changes little between S1 and S2-WET, as the wet year leads to a just 4% increase in non-irrigated wheat in both the South Platte and Arkansas River basins. Similar to corn, the largest changes occur in S3-DRY. Due to the dry year's conditions, nearly a 43% reduction of South Platte non-irrigated wheat area occurs, while the Arkansas River basin non-irrigated wheat disappears completely. The latter basin loses over 283 k hectares of cultivated area. Such large changes in non-irrigated wheat represent expected responses to the drought-like conditions, where yields decline by 26% from the wet year conditions in S2-WET. Therefore, a crop that is dependent on rainfall but not on water via irrigation derived from snowpack and storage will see greater variability in total production as climate changes.

Hay is the third commodity examined in Table 3. The initial decrease in irrigation water in S1 causes a 70 k hectare decline in hay acreage outside the two main basins. After that initial decrease, the hay cultivated in Other Colorado remains constant in S2-WET and S-3-DRY, as that region has sufficient irrigation water, compared to its land resource, and cannot produce other crops competitively. In S3-DRY, irrigated hay increases by 20 k hectares in the Arkansas basin. Overall, the reduction in corn area, due to a substitution into imports, leaves irrigated land available for hay in Arkansas and hence some expansion in hay acreage occurs. In the South Platte, non-irrigated corn and hay, to a lesser extent become competitive on land previously in wheat.

Evaluating production, price and trade effects across climate change simulations. In this section, several important market effects are explained, including the scenarios' effects on total production, trade revenues and prices for major commodities produced in the state. The focus is on climate change effects in S2-WET and S3-DRY, but we consider uncertainties in outcomes and possible alternative scenarios as well.

W*heat.* Wheat consists primarily of non-irrigated production, and is generally exported, with local use equivalent to the level of imports. Production increases in S2-WET by about 436 k tons, or 21%, as more rainfall reaches the crop during its early spring growing season and yields improve by 13%. In S3-DRY, with lower rainfall, non-irrigated wheat area is cut nearly in half, with about 485 k hectares going out of production. The shift towards irrigated corn in the South Platte River basin, noted above, occurs because of a price increase of 10% in S3-DRY. However, the same percentage price increase in wheat does not lead to an increase in non-irrigated production in the Arkansas Valley.

These differing responses between corn and wheat come from varying dependence on imports and the fact that there is no irrigated wheat for the Arkansas River basin in the calibrated model, so that commodity cannot enter even with higher prices.

Thus, the wheat crop is extremely sensitive to how climate change affects rainfall, with the variation in exports between S2-WET and S3-DRY being nearly 1.2 M tons. The actual outcomes will also be affected by the performance of other regions, and, indeed, international supply and demand, as much of Colorado's wheat crop leaves the country. As the Northern Plains outside of Colorado should see greater production of wheat with climate change, downward pressure may be exerted on prices in Colorado, although rising international demand could offset that effect [6]. Higher national and international prices, of course, would reverse some of the decline, as Colorado wheat would remain more competitive than in the scenarios presented here.

In sum, this crop's potential outcomes depend importantly on rainfall variation, as well as the international setting, which affects wheat to a greater degree than other crops. The variability in outlook, however, does not affect other commodities critically, such as corn, hay or cattle, as those are more dependent on irrigation from snowpack and statewide precipitation to a greater extent than the timing and amount of local rainfall.

Cattle Feeding. Cattle feeding is the largest industry in Colorado agriculture and is dependent on selling fattened cattle for slaughter out of the state, although little goes to the international market. In simulations S1 and S2-WET, production declines only slightly from the Base, which is related to an increased cost of feed. However, a higher price exists in the output market, which leads to sales revenues nearly the same as in S1, even though water declines and feed becomes more expensive. On the other hand, in S3-DRY, fed beef production declines by nearly 90 k tons, or 8.4%, due to the significantly higher prices of feed and thus fed beef, which is great enough to dampen demand. The small effect in S2-WET is related to the fact that a quarter of fed beef is sold to consumers in Colorado, where a lower own price elasticity is assumed. Thus, the industry can benefit from increased prices in certain ranges, but higher cost feed eventually makes fed beef less competitive with producers outside the state, particularly in S3-DRY.

Several conflicting trends are not modeled in this research. The first is that increased costs might be incurred for feedlots to adapt to higher temperatures, such as adding sheds and mechanical spraying to protect cattle from heat. Also, the lower quality of hay may require increased quantity in rations. On the other hand, temperatures may increase more in other cattle feeding states, such as Texas, giving Colorado a cost advantage over time. Without knowing which effect will dominate, these variations are left for future work.

Feed sources. Examining changes in feed production highlights overall linkages between products and variations across simulations. From Table 5, it is apparent that corn comprises 85% of overall feed use in the state. That source stays roughly the same until S3-DRY, when irrigated hectares drop due to water shortages, but with high temperatures, yields decline from high heat during pollination. Thus, the quantity of corn used as feed drops by nearly 9% compared to S2-WET.

Table 5. Feed consumed in Base and Climate Change Simulations. Source: Model Runs from Colorado EDMP.

Feed	Base	Simulation			Percentage Change from Base		
		S1 (K tons)	S2-WET (K tons)	S3-DRY (K tons)	S1 (% Change)	S2-WET (% Change)	S3-DRY (% Change)
Hay	3.8	5.0	4.0	4.3	29.7%	5.3%	12.6%
Corn	202.6	201.7	199.8	182.1	−0.5%	−1.4%	−10.1%
Barley	13.8	13.9	14.2	16.7	1.0%	3.0%	20.8%
Oats	2.9	2.9	2.9	2.9	0.0%	0.1%	0.9%
Sorghum	10.4	10.4	10.4	10.4	0.0%	0.0%	0.0%

The use of hay grows from the Base in all three simulations, but source of the forage varies considerably between local production and imports, as is shown in Table 4. The use of hay increases in S1 the most, where the overall water reduction occurs from municipal and industrial uses, rather than due to climatic factors. This is because hay can be imported most easily among the forages, and so there is a swell in imports (which grow by nearly 2.5 times over the Base value). Production drops by 26.1% at the same time, to release irrigation water to be used in other, higher valued crops. In S2-WET, water is less scarce, and yields of non-irrigated pasture and range increase, as do yields of irrigated hay, so less hay is imported and produced.

Production of hay recovers in the third simulation because yield growth of 18% above the Base makes it a profitable user of water. Imports decline because of the general drop in both dairy and cattle feeding seen in S3-DRY. As noted earlier, area is reallocated between the Arkansas and South Platte basins, and the growth occurs due to Colorado feed prices rising in general. In that simulation, corn acreage declines, so irrigated land can shift into hay production. Notably, 283 k hectares are produced in Other Colorado throughout all simulations because there is excess water relative to land in that part of the state.

Corn is the main feed crop that is provided through imports but also has exports. Table 4 showed before that corn is in a net import position, and the internal price does not rise substantially in the first three simulations due to the significance of the import market, where external prices are governed by demand and supply conditions outside Colorado. However, the corn for grain price rises by 10% in S3-DRY due to the general shortage of feed and lower yields of corn in hot and dry conditions. The combination of a water shortage and reduced yields is enough to raise prices to levels where sales of fed beef are affected. This is especially so for exports, which dropped by 9.5% as the industry becomes less competitive. This change leads to lower demand and thus production of corn. Moreover, the ratio of fed beef prices to corn prices declines from about 30 in the first two simulations to 28.7 in S3-DRY, suggesting this change in competitive position.

Effects of Climate Change and Induced Water Loss on Colorado Agricultural Trade. Exports of corn decline by about 22% in S3-DRY relative to the Base scenario, while exports of wheat increase about 19% in S2-WET, due to favorable rainfall and temperature conditions, but decline about 35% in S3-DRY. This leads to a 1.2 M ton swing in exports, which is nearly 60% of average production of wheat in the climate change affected simulations. Beef exports decline about 1.3% and 9.5% in S2-WET and S3-DRY respectively. S2-WET shows 11% decline in wheat imports, while S3-DRY results show that imports of corn, wheat, and dairy increase by 34%, 20% and 104% respectively.

The above changes are all associated with increases in prices, which alter the competitive position of Colorado relative to out of state producers. So, for example, in S3-DRY, wheat prices rise by 10.2% and corn prices increase similarly. For both commodities, exports drop and imports climb as Colorado production becomes more expensive relative to outside sources. Imports of Hay increase in the simulations, with hay imports more than tripling in value in S1 relative to the Base. In S3-DRY, less corn is grown with the reduction in cattle feeding, and thus irrigated land becomes available for hay, which expands from higher prices. This latter outcome is related to the assumption that yields increase for hay from the longer growing season, but decrease in corn from heat and rainfall variation.

Table 6 gives an important perspective on model outcomes provided above. The import and export elasticities for major commodities are first presented, which were constructed to reflect differing external positions. These are key assumptions, of course, because they have a large effect on quantity and price changes in a given simulation. The values are all high, so a "5", for example, indicates that a 1% change in price will lead to a 5% change in quantity, implying quite a large response. Thus, the exports of wheat and fed beef are very responsive to how the internal price changes with respect to the import or export price, which is consistent with a small open economy where local industries face much competition from external sources of supply.

Table 6. Export and import elasticities in the Colorado EDMP, and trade proportions for key commodities.

Commodity	Elasticities	Export or import percent of production
Corn exports	2	15.40%
Wheat exports	5	99.60%
Fed beef exports	5	75.50%
Corn imports	3	90.80%
Hay imports	2	62.70%
Wheat imports	3	11.60%

Corn and wheat's import and export elasticities are worthy of specific mention. The wheat export elasticities exceed its import elasticities, capturing the reality that marketing and distribution systems are export oriented, and there will be a tendency to export wheat output. Wheat production is less likely to develop domestic uses that require more imports, and thus that elasticity is somewhat lower. The reverse is true for corn, where imports support a large feeding industry and a projected ethanol industry, so the import elasticity is higher than the export elasticity.

The wheat import elasticity is lower than the export elasticity to take into account the fact that Colorado is a surplus producer, and, therefore, most infrastructure and institutional relationships focus on exports rather than increasing imports. However, both wheat and corn imports are still elastic relations, as many users of corn and wheat in the Eastern Plains, especially, can purchase needed quantities from nearby locations in Kansas and Nebraska, so it is easy to obtain imports and thus these relationships should be elastic.

The hay elasticity for imports is lower due to an assumption of significant transport costs and therefore tighter regional markets. To bring in more imports to Colorado, therefore, prices must rise faster than in the more widely traded corn and wheat markets. This has a fairly large effect on the local market in S3-DRY, where prices rise internally, forage use is cut, and dairy production decreases. The higher internal prices, driven partly by this elasticity assumption, leads to growth in hay production on irrigated hectares in Arkansas in S3-DRY, especially as corn production declines due to lower demand.

Imports and exports play an important role in describing climate change impacts on Colorado. Exports of wheat and beef, and imports of corn, are all greater than 90% of domestic production, so these products are clearly dependent on external economic performance and trends. We noted earlier that almost all wheat produced in Colorado leaves the state, often for international destinations. The large beef feeding industry is export-oriented, with about three quarters of production leaving the state. Hay is also a commodity where the import market is used quite variably across the simulations.

Welfare Effects. Because the model captures changes in prices and quantities, and has demand and supply functions embedded in the objective function, it is possible to determine changes in producer and consumer surpluses under the different simulations. In this fashion, the model shows how costs of climate change are borne, and could be employed to assess the value of various mitigation strategies in a future study. These results are presented in Figure 2. The measures of economic surplus show approximately a $10.7 M reduction in the S3-DRY scenario, compared to about $2.7 M in the wet year in S2-WET. In other words, the agricultural economy in Colorado loses nearly five times as much in a dry year climate relative to a wet year. The S1 climate scenario is predicted to produce economic net welfare impact that fits in the middle between S2-WET and S3-DRY (at about $6.2 M).

Figure 2. Changes in Producer Surplus (PS) and Consumer Surplus (CS), Million of Dollars.

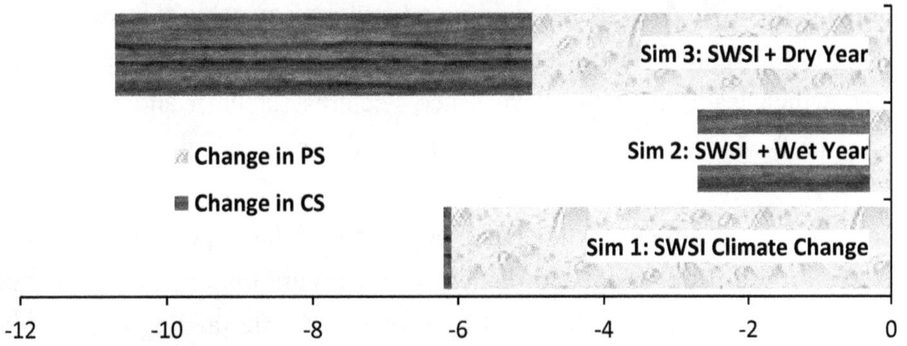

	Sim 1: SWSI Climate Change	Sim 2: SWSI + Wet Year	Sim 3: SWSI + Dry Year
Change in PS	-6.1	-0.3	-5
Change in CS	-0.1	-2.4	-5.7

In S1, most impacts fall on producers through reduced hay area, which has the greatest effect due to its water use, and which is made up by added imports and reduced dairy production. The largest effects naturally come in the dry year simulation, where cultivated area is reduced by up to 60% for some crops and yields can decline by over 10%. The total losses in S3 of more than $10 M are split about evenly between consumers and producers. Even though prices for livestock and major crops often increase by up to 10%, the decline in quantities offsets those better prices, and there is a net loss in producer surplus, which occurs because of the openness of the agricultural economy. The consumers lose in S3-DRY due to the higher overall prices.

Conclusions

Using an Economic Displacement Mathematical Programming (EDMP) model, derived from Harrington and Dubman [34] of the USDA's Economic Research Service. This study examines the effects of climate change on agriculture in Colorado taking into account of selected features projected several decades into the future. Initially, an overview of agriculture in the state and its dependence on water, a critical input, is described. The overview shows that the agricultural economy in Colorado is dominated by livestock, which accounts for 67% of total receipts. Crops, including feed grains and forages, account for 33% of production. Most of agriculture is based on irrigated production, which depends on both groundwater, especially from the Ogallala aquifer, and surface water that comes from runoff derived from snowpack in the Rocky Mountains. Climate studies point to decline in runoff from 6% to 20% by 2050. The timing of runoff is projected to begin and peak earlier in the spring and late-summer, and overall flows may be reduced.

The climate change scenarios evaluated in this paper include three simulations relative to a Base scenario that reflects some key characteristics with regard to future water and yield effects of climate change. Following SWSI projections, the base reflects demographics and economic changes from the calibrated model for 2007. The Base scenario models a 10.3% reduction in agricultural water from increased municipal and industrial water demand, and assumes a 75% increase in corn extracted-ethanol production. The first simulation reduces agricultural water availability by a further 14.0%, for a combined decrease of 24.3%, due to climatic factors and related groundwater depletion. The second simulation describes a year with warmer than historical average temperatures and wetter conditions, which negatively affect yields of irrigated corn and milking cows, but it improves yields for non-irrigated wheat, corn silage, irrigated hay, rangeland and pasture. In contrast, the last simulation describes a drought year, which leads to reduced harvested hectares for corn and wheat, and negatively affects yields for dry land wheat, irrigated corn, pasture and rangeland, while irrigated corn silage and hay output increase.

Three commodities examined in this paper account for a large percent of production in the Colorado agricultural sector: fed beef, wheat and dairy; two others are major sources of feed, including hay and corn. All are strongly affected by the S3-DRY scenario. Cattle feeding is dependent on exports out of the state, and in S3-DRY, fed beef production declines by 7.5% due to the significantly higher prices of feed and the resulting effect on output price. For corn, the hectares decrease by about 38% on irrigated land in both regions, while in the Arkansas basin, non-irrigated land declines by 75%. Due to the dry year's conditions, nearly a 50% reduction of South Platte non-irrigated wheat area occurs, while the Arkansas River basin non-irrigated wheat disappears completely. The wheat crop is extremely sensitive to how rainfall is affected by climate change, with the variation in exports being nearly 1.5 M tons.

The dairy sector reacts strongly to climate variation, given that production decreases by 18% in both warmer scenarios. Dairy is the second largest user of hay, after cow calf producers, and it is the second largest user of grain, after cattle feeding, as its rations require more of each basic feedstuff. Therefore, as feed shortages develop, dairy declines first and frees up significant proportions of grain and forage. The reduction in corn area leaves irrigated land available for hay production in the Arkansas basin, and expansion in irrigated hay occurs in the same basin in drought scenario. In the South Platte, non-irrigated corn becomes competitive on the land that was previously in wheat. Notably, 280 k hectares are in hay production in other parts of Colorado throughout all simulations because excess water relative to land exists in that part of the state.

This model has not taken into account farmers' adaptation strategies, which would reduce the climate impact on yields. Such strategies might include changing planting schedules, production practices or technologies, and the introduction of drought-tolerant varieties. Also, the model has not reflected climate-induced shifts in planting decisions and production practices that lead to various environmental impacts and higher costs. There could be soil and water quality effects through nutrient loss and soil erosion, and a greater use of pesticides to combat a higher prevalence of pests.

These environmental dimensions can be fruitful areas to examine in future research, as would be the development of a wider range of conditions in the analysis of climate change effects in the future. Some of the latter areas could be to look at various productivity growth scenarios before adding the effects of climate change, and also broader alternatives in performance of different commodities. This paper assumes certain large effects, such as the increase in yields for hay and the decrease in dairy output, but others, such as using the current set of relative prices and import and export positions as starting points, may seem to understate the climate change impacts on the agricultural economy of Colorado. A more extensive examination of these settings could provide additional insights.

Acknowledgment

The authors thank David Harrington and Robert Dubman at the Economic Research Service, U.S. Department of Agriculture, for providing the U.S. Equilibrium Displacement Mathematical Programming Model, which the authors modified to reflect Colorado's agricultural economy and water specifications.

Appendix

Positive Mathematical Programming

Returning to the matrix notation of Equation (1), Z is subject to the following constraints:

$A_{11}x \leq$ free Indicator accounts, necessary for analytical purposes, not shown in the Tableau;

$A_{21}x \leq b$ Resource constraints;

$A_{31}x \leq 0$ Commodity balance equations;

$I31x = c$ Calibration constraints, dropped after calibration;

$U11 \leq 0$ Input accounts.

The additional notation is:

A is the matrix of technical coefficients;

I is an identity matrix of calibration constraints;

U is the matrix of inputs in dollar value to sector's activities;

b is a vector of right hand sides of resource constraints;

c is a vector of calibration quantity targets used only in calibration phase.

The resource constraints involve land and water for crop activities. Cropland, pasture, range land and land in the conservation reserve programs are quantified and include land fallowed as part of crop rotations including wheat-fallow. The supply of water available to agriculture is fixed, while the demand for water is exogenously determined for each crop by the State of Colorado's Consumptive Use Model (StateCU) component of the CDSS, which is based on a modified Blaney-Criddle method. (Other constraints include livestock facilities for livestock and labor for both crop and livestock activities).

The block of commodity balance equations runs across the production, demand and trade sections of the model. These are accounting constraints that distribute production across its uses. Corn, wheat and hay production are separated by location for the South Platte, Arkansas, and St. Luis Valley, and the Upper Colorado basins, and are identified by whether they are irrigated or non-irrigated production. Within this block, the two rows for corn and ethanol/distilled grain are highlighted. The corn balance equation allocates crop production from each basin and type of farming activity (irrigated *versus* non-irrigated) across basins and imports to ethanol production, domestic non-farm sales, and exports. In addition feed use of corn is calculated as a residual and transferred to the grain ration equation.

For example, the following is the corn commodity balance equation with the variable acronyms:

$$-60.818 \text{ SPCRND} - 176.868 \text{ IRSPCRN} - 179.625 \text{ ARCRNIR} - 45.75 \text{ DCRNAR} - 134.134 \text{ CORNCO} + \text{CRNTUS} + 0.357 \text{ ETHCO} + \text{SELCRNCO} + \text{EXPCRNCO} - \text{IMPCRNCO} \leq 0 \tag{A1}$$

where, SPCRND, IRSPCRN, ARCRNIR, DCRNAR and CORNCO are corn production activities (harvested hectares) in South Platte non-irrigated, irrigated land, Arkansas non-irrigated, Arkansas irrigated land and the rest of Colorado; CRNTUS is the production allocated to feed; ETHCO is the ethanol production in M gallons; SELCRNCO, EXPCRNCO and IMPCRNCO are the levels of non-farm domestic sales, exports and imports in tons. The coefficients on the hectares are yields (tons/hectare), while the coefficient with ethanol production is the conversion ratio (liters of ethanol/ton of corn).

The feed requirements are calculated as intermediate inputs and are not priced in CDEMP. The model includes two rations. The grain ration equation is formulated as follows:

$$\sum_i \sum_b \alpha'' g - .064 \, \text{eth} + \sum \beta'' K \leq 0 \qquad (A2)$$

where, α is the vector of coefficients converting crops into feed ration components; and g is the vector of grain feed crops (corn, barley, oat and sorghum); eth is the level of ethanol production; and β is the vector of ration requirements in as fed form by livestock types; and k is the vector of livestock activity levels. Note that both g and K are subsets of cl_{ib}, and α and β are subsets of A_{31}, the matrix of technical coefficients.

The forage ration equation has similar structure:

$$\sum_i \sum_b \mu''' h + \sum \theta''' k \leq 0 \qquad (A3)$$

where, μ is the vector of coefficients converting hay and pasture forage into feed ration components and h is the vector of forage activities (silage, cropped hay and pastures, permanent pastures and rangeland), θ is the vector of ration requirements in *as fed* form identified by livestock types k. Here h and k are both subsets of cl_{ib}, and μ and θ are subsets of A_{31}.

Harrington and Dubman [35] suggested changing one or more of the following EDMP model's parameter(s) to calibrate a base scenario:

1- Modify the scenario intercept for parallel shift of supply or demand function;
2- Modify the Hessian for rotation of the supply or demand function;
3- Modify the Right Hand Side (RHS) coefficients to change the resource availability;
4- Change the crop's yield, livestock productivity, or change the transfer from primary to semi or finished product coefficients.

Table A1. Area and production of crops and livestock activities, actual, and calibrated values for the Colorado EDMP.

Crop or Commodity	Units	Historical 2007 quantity	Calibrated quantity	Hessian element	Intercept
Ethanol	Million Liters	648.97	660.96	0.00	1.51
South platte dry corn	Million Hectares	0.10	0.11	−76.73	909.07
South platte irrigated corn	Million Hectares	0.21	0.23	−38.93	1173.13
Arkansas dry corn	Million Hectares	0.04	0.13	−81.46	436.30
Total corn	**Million Hectares**	**0.36**	**0.47**		
South Platte dry wheat	Million Hectares	0.04	0.00	−183.25	1160.79
South Platte irrigated wheat	Million Hectares	0.55	0.63	−11.25	1549.58
Arkansas dry wheat	Million Hectares	0.33	0.31	−22.30	703.28
Wheat, other [a]	Million Hectares	0.04	0.00	−157.75	2262.93
Total wheat	**Million Hectares**	**0.87**	**0.93**		
Sorghum	Million Hectares	0.07	0.07	−44.27	1683.21
Potatoes	Million Hectares	0.02	0.03	−13871.39	4041.04

Table A1. *Cont.*

Crop or Commodity	Units	Historical 2007 quantity	Calibrated quantity	Hessian element	Intercept
South Platte irrigated. hay, all	Million Hectares	0.15	0.12	−10.04	1077.62
Arkansas dry hay, all	Million Hectares	0.09	0.11	−121.73	1082.96
Hay all, other	Million Hectares	0.31	0.34	−9.71	3251.96
Hay all, total	**Million Hectares**	**0.55**	**0.57**		
Fed beef	Thousand Ton	1235.6	1241.9	−0.2	182.1
Hogs,	Thousand Ton	161.6	165.7	−57.3	76.4
Dairy	Thousand Ton	1228.3	1236.5	−1.1	42.3
Broiler	Thousand Ton	157.5	173.4	−6.1	92.8
Eggs, independent	Million dozens	8.83	9.72	0	1.9
Eggs, contracted	Million dozens	79.5	87.45	0	2.5
Turkey, independent	Thousand Ton	13.6	16.8	−108.3	108
Turkey, contracted	Thousand Ton	20.9	21.3	−141.3	135.4

Note: [a] Other basins include San Luis Valley and Colorado River basin.

References

1. Colorado Water Conservation Board. *Statewide Water Supply Initiative Report Overview*; Colorado Department of Natural Resources: Denver, CO, USA, 2004.

2. Ray, A.; Barsugli, J.; Averyt, K.; Wolter, K.; Hoerling, M.; Doesken, N.; Udall, B.; Webb, R.S. *Climate Change in Colorado: A Synthesis to Support Water Resources Management and Adaptation*; University of Colorado Boulder: Boulder, CO, USA, 2008.

3. Parry, M.L. *Climate Change 2007: Impacts, Adaptation and Vulnerability: Contribution of Working Group II to the Fourth Assessment Report of the Intergovernmental Panel on Climate Change*; Cambridge University Press: Cambridge, UK, 2007; Volume 4.

4. Solomon, S.D.; Qin, M.; Manning, Z.; Chen, M.; Marquis, K.B. *Contribution of Working Group I to the Fourth Assessment Report of the Intergovernmental Panel on Climate Change*; Cambridge University Press: Cambridge, UK, 2007.

5. Thorvaldson, J.; Pritchett, J. *Economic Impact Analysis of Irrigated in Four River Basins in Colorado*; Colorado Water Resources Research Institute: Fort Collins, CO, USA, 2006.

6. Gunter, A.; Goemans, C.; Pritchett, J.G.; Thilmany, D.D. Linking an Equilibrium Displacement Mathematical Programming Model and an Input-Output Model to Estimate the Impacts of Drought: An Application to Southeast Colorado. In Proceedings of Agricultural & applied Economics Association's 2012 AAEA Annual Meeting, 12–14 August 2012; Agricultural and Applied Economics Association: Seattle, WA, USA, 2012.

7. Malcolm, S.; Marshall, E.; Aillery, M.; Heisey, P.; Livingston, M.; Day-Rubenstein, K. *Agricultural Adaptation to a Changing Climate: Economic and Environmental Implications Vary by US Region*; USDA-ERS Economic Research Report No. 136; United States Department of Agriculture-Economic Research Service (USDA-ERS): Washington, DC, USA, 2012.

8. Islam, A.; Ahuja, L.R.; Garcia, L.A.; Ma, L.; Saseendran, A.S.; Trout, T.J., Modeling the impacts of climate change on irrigated corn production in the Central Great Plains. *Agric. Water Manag.* **2012**, *110*, 94–108.

9. Herrero, M.P.; Johnson, R. High temperature stress and pollen viability of maize. *Crop. Sci.* **1980**, *20*, 796–800.

10. Kelly, E.Z.; Tunc-Ozdemir, M.; Harper, J.F. Temperature stress and plant sexual reproduction: Uncovering the weakest links. *J. Exp. Bot.* **2010**, *61*, 1959–1968.

11. Izaurralde, R.C.; Thomson, A.M.; Morgan, J.; Fay, P.; Polley, H.; Hatfield, J.L. Climate impacts on agriculture: Implications for forage and rangeland production. *Agron. J.* **2011**, *103*, 371–381.

12. Morgan, J.A.; LeCain, D.R.; Pendall, E.; Blumenthal, D.M.; Kimball, B.A.; Carrillo, Y.; Williams, D.G.; Heisler-White, J.; Dijkstra, F.A.; West, M. C4 grasses prosper as carbon dioxide eliminates desiccation in warmed semi-arid grassland. *Nature* **2011**, *476*, 202–205.

13. Backlund, P.; Janetos, A.; Schimel, D.; Walsh, M. The effects of climate change on agriculture, land resources, water resources, and biodiversity in the United States. In *The Effects of Climate Change on Agriculture, Land Resources, Water Resources, and Biodiversity in the United States*; Synthesis and Assessment Report 4.3; U.S. Department of Agriculture: Washington, DC, USA, 2008.

14. Preston, B.L.; Jones, R. *Climate Change Impacts on Australia and the Benefits of Early Action to Reduce Global Greenhouse Gas Emissions*; The Commonwealth Scientific and Industrial Research Organisation (CSIRO): Clayton South, Australia, 2006.

15. Moons, C.P. H.; Sonck, B.; Tuyttens, F.A.M. Importance of outdoor shelter for cattle in temperate climates. *Livest. Sci.* **2014**, *159*, 87–101.

16. Lambertz, C.; Sanker, C.; Gauly, M. Climatic effects on milk production traits and somatic cell score in lactating Holstein-Friesian cows in different housing systems. *J. Dairy Sci.* **2014**, *97*, 319–329.

17. National Agricultural Statistics Service (NASS). Department of Agriculture. Census of Agriculture. Available online: http://www.agcensus.usda.gov/ (accessed on 15 June 2012).

18. Colorado Water Conservation Board. *Conservation Levels Analysis Final Report*; Colorado Department of Natural Resources: Denver, CO, USA, 2010.

19. Hardling, B.L; Wood, A.W.; Prairie, J.R. The implications of climate change scenario selection for future stream flow projection in the Upper Colorado River Basin. *Hydrol. Earth Syst. Sci. Discuss.* **2012**, *9*, 847–894.

20. Howitt, R.E.; MacEwan, D.; Medellín-Azuara, J.; Lund, J.R. *Economic Modeling of Agriculture and Water in California Using the Statewide Agricultural Production Model*; Department of Agricultural and Resource Economics, Department of Civil and Environmental Engineering, Center for Watershed Sciences: Davis, CA, USA, 2010.

21. Preckel, P.V.; Harrington, D.; Dubman, R. Primal/dual positive math programming: Illustrated through an evaluation of the impacts of market resistance to genetically modified grains. *Am. J. Agric. Econ.* **2002**, *84*, 679–690.

22. Schmid, E.; Sinabell, F. *Using the Positive Mathematical Programming Method to Calibrate Linear Programming Models*; University für Bodenkultur Wien, Department für Wirtschafts-u. Sozialwiss, Inst. für NachhaltigeWirtschaftsentwicklung: Vienna, Austria, 2005.

23. Heckelei, T.; Britz, W. Models Based on Positive Mathematical Programming: State of the Art and Further Extensions. In *Modeling Agricultural Policies: State of the Art and New Challenges*; Monte Università Parma: Parma, Italy, 2005; pp. 48–73.

24. De Frahan, B.H.; Buysse, J.; Polomé, P.; Fernagut, B.; Harmignie, O.; Lauwers, L.; Van Huylenbroeck, G.; Van Meensel, J. Positive Mathematical Programming for Agricultural and Environmental Policy Analysis: Review and Practice. In *Handbook of Operations Research in Natural Resources*; Springer: New York, NY, USA, 2007; pp. 129–154.

25. Buysse, J.; Van Huylenbroeck, G.; Lauwers, L. Normative, positive and econometric mathematical programming as tools for incorporation of multifunctionality in agricultural policy modeling. *Agric. Ecosyst. Environ. Ecosyst. Environ.* **2007**, *120*, 70–81.

26. Heckelei, T.; Wolff, H. Estimation of constrained optimisation models for agricultural supply analysis based on generalised maximum entropy. *Eur. Review Agric. Econ.* **2003**, *30*, 27–50.

27. Osterburg, B.; Offermann, F.; Kleinhanss, W. A Sector Consistent Farm Group Model for German Agriculture. In *Agricultural Sector Modeling and Policy Information Systems*; *Vauk Verlag*: Kiel, Germay, 2001; pp. 152–160.

28. Judez, L.; De Miguel, J.; Mas, J.; Bru, R. Modeling crop regional production using positive mathematical programming. *Math. Comput. Model.* **2002**, *35*, 77–86.

29. Baskaqui, A.; Butault, J.; Rousselle, J. Positive Mathematical Programming and Agricultural Supply within EU under Agenda 2000. In Proceedings of the 65th European Seminar of the European Association of Agricultural Economists (EAAE), Bonn, Germany, 29–31 March 2000, Wissenschaftsverlag Vauk Kiel KG: Kiel, Germany, 2001; p. 200.

30. Paris, Q.; Montresor, E.; Arfini, F.; Mazzocchi, M.; Heckelei, T.; Witzke, H.; Henrichsmeyer, W. An Integrated Multi-Phase Model for Evaluating Agricultural Policies through Positive Information, Agricultural Sector Modelling and Policy Information Systems. In Proceedings of the 65th European Seminar of the European Association of Agricultural Economists (EAAE), Bonn, Germany, 29–31 March 2000, Wissenschaftsverlag Vauk Kiel KG: Kiel, Germany, 2001; pp. 100–110.

31. Sinabell, F.; Streicher, G. Programme Evaluation with Micro-Data: The Use of FADN Data to Evaluate Effects on the Market Situation of Programme Participants. In Proceedings of 87th EAAE-Seminar: Assessing Rural Development Policies of the CAP, Vienna, Austria, 21–23 April 2004.

32. Howitt, R.E. Positive mathematical programming. *Am. J. Agric. Econ.* **1995**, *77*, 329–342.

33. Muth, R.F. The derived demand curve for a productive factor and the industry supply curve. *Oxf. Econ. Pap.* **1964**, *16*, 221–234.

34. Piggott, R.R.; Piggott, N.E.; Wright, V.E. Approximating farm-level returns to incremental advertising expenditure: Methods and an application to the Australian meat industry. *Am. J. Agric. Econ.* **1995**, *77*, 497–511.

35. Samuelson, P.A. Spatial price equilibrium and linear programming. *Am. Econ. Review* **1952**, *42*, 282–303.

36. Harrington, D.H.; Dubman, R. *Equilibrium Displacement Mathematical Programming Models: Methodology and a Model of the U.S. Agricultural Sector*; Technical Bulletin No. (TB-1918); United States Department of Agriculture: Washington, DC, USA, 2008.

37. Johnson, G.; Quance, C.L. *The Overproduction Trap in US Agriculture: A Study of Resource Allocation from World War I to the Late 1960's*; The John Hopkins University Press: Baltimore, MD, USA, 2011

38. Connor, J.D.; Schwabe, K.; King, D.; Knapp, K. Irrigated agriculture and climate change: The influence of water supply variability and salinity on adaptation. *Ecol. Econ.* **2012**, *77*, 149–157.

39. Henseler, M.; Wirsig, A.; Herrmann, S.; Krimly, T.; Dabbert, S. Modeling the impact of global change on regional agricultural land use through an activity-based non-linear programming approach. *Agric. Syst.* **2009**, *100*, 31–42.

40. Qureshi, M.E.; Ahmad, M.-U.-D.; Whitten, S.M.; Kirby, M. A Multi-Period Positive Mathematical Programming Approach for Assessing Economic Impact of Drought in the Murray-Darling Basin, Australia. In Proceedings of Australian Agricultural and Resource Economics Society, 56th Conference, Freemantle, Australia, 7–10 February, 2012.

41. Withey, P.; van Kooten, G.C. *The Effect of Climate Change on Land Use and Wetlands Conservation in Western Canada: An. Application of Positive Mathematical Programming*; Working Paper 2011–04; Resource Economics and Policy Analysis Research Group (REPA), Department of Economics, University of Victoria: Victoria, Canada, 2011.

42. Schlenker, W.; Hanemann, W.M.; Fisher, A.C. Water availability, degree days, and the potential impact of climate change on irrigated agriculture in California. *Clim. Change* **2007**, *81*, 19–38.

43. Bauman, A.; Goemans, C.; Pritchett, J.; McFadden, D.T. Estimating the economic and social impacts from the drought in Southern Colorado. *J. Contemp. Water Res. Educ.* **2013**, *151*, 61–69.

44. Schaible, G.D.; Kim, C.; Aillery, M.P. Dynamic adjustment of irrigation technology/water management in western US agriculture: Toward a sustainable future. *Can. J. Agric. Econ.* **2010**, *58*, 433–461.

45. Pritchett, J.G.; Davies, S.P.; Fathelrahman, E.; Davies, A. Welfare Impacts of Rural to Urban Water Transfers: An Equilibrium Displacement Approach. In Proceedings of Agricultural & Applied Economics Association (AAEA), Canadian Agricultural Economics Society (CAES), & Western Agricultural Economics Association (WAEA) Joint Annual Meeting, Denver, CO, USA, 25–27 July 2010.

46. Colorado Decision Support System (CDSS). Colorado Division of Water Resources. Databases and software. Available online: http://cdss.state.co.us/Pages/CDSSHome.aspx (accessed on 20 March 2012).

47. Takayama, T.; Judge, G.G. Equilibrium among spatially separated markets: A reformulation. *Econom. J. Econ. Soc.* **1964**, *32*, 510–524.

48. Helming, J.F. A Model of Dutch Agriculture Based on Positive Mathematical Programming with Regional and Environmental Applications. Ph.D. Thesis. Wageningen University, Wageningen, the Netherlands, 11 February 2005.

49. Colorado Water Conservation Board. *Colorado River Water Availability Study, March, 2010. Phase I. Draft*; Colorado Department of Natural Resources: Denver, CO, USA, 2010.

9

Potential Impacts of Climate Change on Precipitation over Lake Victoria, East Africa, in the 21st Century

Mary Akurut [1,2,*], **Patrick Willems** [1,3,*] **and Charles B. Niwagaba** [2]

[1]   Department of Civil Engineering, KU Leuven, Kasteelpark Arenberg 40, bus 2448, Leuven 3001, Belgium; E-Mail: patrick.willems@bwk.kuleuven.be

[2]   Department of Civil and Environmental Engineering, Makerere University Kampala, P. O. Box 7062, Kampala 00256, Uganda; E-Mail: cniwagaba@cedat.mak.ac.ug

[3]   Department of Hydrology and Hydraulic Engineering, Vrije Universiteit Brussel, Pleinlaan 2, Brussels 1050, Belgium

*   Author to whom correspondence should be addressed; E-Mail: mary.akurut@bwk.kuleuven.be;

Abstract: Precipitation over Lake Victoria in East Africa greatly influences its water balance. Over 30 million people rely on Lake Victoria for food, potable water, hydropower and transport. Projecting precipitation changes over the lake is vital in dealing with climate change impacts. The past and future precipitation over the lake were assessed using 42 model runs obtained from 26 General Circulation Models (GCMs) of the newest generation in the Coupled Model Intercomparison Project (CMIP5). Two CMIP5 scenarios defined by Representative Concentration Pathways (RCP), namely RCP4.5 and RCP8.5, were used to explore climate change impacts. The daily precipitation over Lake Victoria for the period 1962–2002 was compared with future projections for the 2040s and 2075s. The ability of GCMs to project daily, monthly and annual precipitation over the lake was evaluated based on the mean error, root mean square error and the frequency of occurrence of extreme precipitation. Higher resolution models (grid size <1.5°) simulated monthly variations better than low resolution models (grid size >2.5°). The total annual precipitation is expected to increase by less than 10% for the RCP4.5 scenario and less than 20% for the RCP8.5 scenario over the 21st century, despite the higher (up to 40%) increase in extreme daily intensities.

Keywords: climate change; precipitation; general circulation models (GCMs); representative concentration pathways (RCP); Lake Victoria

---

1. Introduction

Lake Victoria, Africa's largest fresh water lake covers a surface area of about 68,800 km^2 shared across three East African countries: Uganda (45%), Kenya (6%), and Tanzania (49%). Over 30 million inhabitants depend on Lake Victoria for their livelihoods. Therefore, precipitation changes over the lake are likely to affect the quality of life of many within the East Africa region. Lake Victoria has a complex shoreline structure comprising gulfs and bays that provide potable water abstraction points and also receive municipal and industrial waste from adjacent urban centers.

Due to the vast size of the Lake Victoria basin, it is considered that the average annual lake precipitation almost balances the annual evapotranspiration. Therefore, precipitation variations significantly influence water levels in Lake Victoria. This notion has been applied by several authors to study the water balance of the lake, often translated as changes in the lake levels or outflow regimes—with most variations in the water balance being attributed to the different calculation periods and methods used in estimation of the different balance components *i.e.*, evapotranspiration, inflows, outflows and precipitation [1–5]. About 80% of the Lake Victoria refill is predominantly precipitation compared to the 20% from basin discharge [6]. Satellite remote sensing data was applied in [7] to monitor the water balance of Lake Victoria in comparison to other water bodies in the vicinity—climate forcing explained half of the lake level trends while the outflow patterns were responsible for the other half. Climate forcing is generally affected by the amount of aerosols and greenhouse gases (GHG) in the atmosphere. GHGs absorb and re-emit energy radiated from the Earth's surface, leading to a warming or cooling effect and changes in the Earth's energy balance with time. Increasing greenhouse gas concentrations in the atmosphere leads to warming which in turn causes global atmospheric water vapor and precipitation to increase. Aerosols directly absorb and scatter incoming solar radiation leading to cooling at the surface and a reduction in precipitation. They can also affect precipitation through complex interactions with clouds [8]. At regional scales, changes in precipitation can also be influenced by anthropogenic activities that affect atmospheric transport of water vapor and circulation changes.

The importance of global precipitation changes as addresssed in [8] by the Intergovernmental Panel on climate Change (IPCC) fifth Assessment Report (AR5) suggests a need to understand and project effects of extreme climate conditions. This paper evaluates the newest generation models used in the CMIP5 project with the purpose of studying impact of climate change on the quantity and quality of water in Lake Victoria. Precipitation was aggregated at different temporal scales; daily, monthly and annually. Model evaluation was based on a range of statistical measures and visual graphical comparison for the same aggregation periods in order to postulate possible precipitation changes over Lake Victoria.

2. Data and Methods

2.1. Description of the Study Area

Lake Victoria is located in the upper Nile basin in East Africa within latitudes 00°30′00″ N to 03°00′00″ S and longitudes 31°30′00″ E to 35°00′00″ E. The Lake surface is at an average elevation of about 1135 m.a.s.l (Figure 1). Lake Victoria covers a total catchment area of about 258,000 km². The lake itself contributes about 27% of the total catchment area. Generally, the Lake Victoria basin climate is characterized by substantial precipitation occurring throughout the year; however, there are two distinct rainy seasons in which monthly precipitation is generally greater than 10% of the average monthly precipitation. Heavier precipitation occurs in the March-April-May (MAM) season, while the longer rainy season occurs in October-November-December (OND). Climate variability for the lake basin region is influenced by both large-scale and meso-scale circulations resulting from complex interactions of the Inter-Tropical Convergence Zone (ITCZ) and El Nino Southern Oscillation (ENSO), Quasi-biennial Oscillations, large-scale monsoonal winds, and extra-tropical weather systems [9–12].

Figure 1. (**a**) Location of Lake Victoria within Africa; (**b**) Coordinates where general circulation model (GCM) precipitation output for Lake Victoria was extracted.

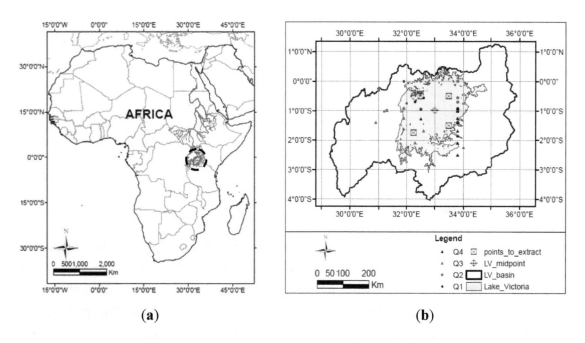

(**a**)   (**b**)

2.2. Precipitation and Lake Levels

Precipitation over Lake Victoria experienced a predominantly positive trend over the 20th century [12]. A sharp increase in water levels occurred in 1962—it was mainly attributed to the high precipitation in that year and the related high tributary inflows [3,13]. Precipitation occurrence had the largest effect on the lake levels and flow exiting the basin at the Victoria Nile river, while irrigation and hydropower developments had modest effects on these levels and flows [14]. However, commissioning of the Owen Falls Dam, located on the White Nile in 1954 (just prior to the lake rise in 1962) could also have had an impact on the water levels as the lake regained its level as noted by [13,15]. Figure 2 shows the cumulative precipitation and discharge trends from the Lake Victoria catchments compared to the water levels in the lake. It can be deduced that tributary inflows were more significant in increasing lake levels in 1962 and 1998, which years coincided with the El Nino years [9,10] as depicted by the jumps in the cumulative tributary inflows. In conclusion, both human management roles and natural factors affected the lake levels, but precipitation clearly is the major factor. Climate change impact investigations on the Lake Victoria water levels therefore should focus on the future changes in precipitation.

Figure 2. Lake level variations over time compared to cumulative average precipitation over the lake and cumulative total inflow into the lake. The calculated lake level variations are based on the precipitation and inflow [16].

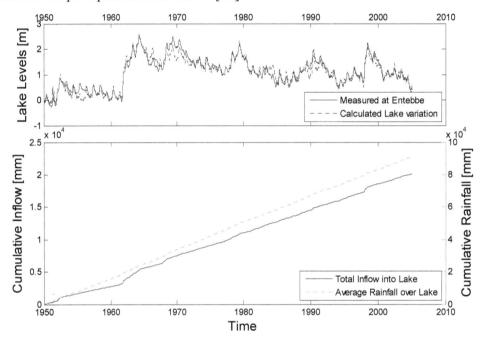

2.3. Climate Model Simulations

General circulation models (GCMs) are numerical models that describe physical processes of the global climate system in the atmosphere, ocean, cryosphere and land surface in response to changing GHG and aerosol concentrations. GCMs provide geographical and physical estimates of regional climate and climate change using three dimensional grids over the globe. The newest generation GCMs are used in the CMIP5 to understand the past and future climate changes. These are the models upon which the recent Fifth Assessment Report (AR5) of the IPCC is based [8]. The CMIPs attempt to address major priorities and incorporate ideas from a wide range of climate modelling communities. The climate change modelling experiments are integrated using atmosphere-ocean global climate models (AOGCMs). These models respond to specified, time-varying concentrations of various atmospheric constituents e.g., GHGs and include interactive representation of the atmosphere, ocean, land and sea ice. CMIP5 also introduces coupling of biogeochemical components to account for closing of carbon fluxes between the oceans, atmosphere and terrestrial biosphere carbon reservoirs for long term simulations in the earth system models. They are capable of using time-evolving emissions of constituents to interactively compute concentrations.

The main difference between these CMIP5 projections and the previous CMIP projections is that their climate change projections include policy intervention and mitigation measures [17]. CMIP5 provides a large set of runs that enable systematic model inter-comparison within each type of experiment and credible multi-model analysis. The core experiments include the historical runs covering much of the industrial period (mid-19th century to the near-present) and future projection simulations forced with specific GHG concentrations and anthropogenic aerosols emissions dubbed "Representative Concentration Pathways" (RCPs) e.g., RCP4.5 and RCP8.5. RCP8.5 is consistent with the high emissions scenario in which the radiative forcing increases throughout the 21st century before reaching 8.5 Wm^{-2} at the end of the century, while RCP4.5 signifies a mid-range mitigations emissions scenario where GHG valuation policies are applied to stabilize atmospheric radiative forcing to 4.5 Wm^{-2} in 2100 (Figure 3). These two CMIP5 scenarios were considered in this study as a basis of exploring climate change impacts and policy issues. RCPs enable investigations of uncertainties related to carbon cycle and atmospheric chemistry. They span a wide range of total forcing values though they do not cover the full range of emissions in the literature, particularly for aerosols [8].

Figure 3. Representative concentration pathways. (**a**) Changes in radiative forcing relative to pre-industrial conditions; (**b**) Energy and industry CO_2 emissions for the different representative concentration pathway (RCP) candidates. The range of emissions in the recent (post 2001) literature is presented as a thick dashed curve for the maximum and minimum while the shaded area represents the 10th to 90th percentiles [17].

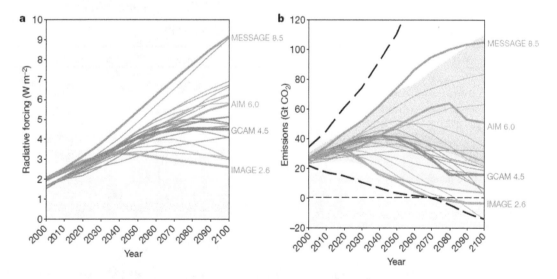

According to [17], a realistic climate model should exhibit internal variability with spatial and temporal structure like the observed. However, in the long-term simulations, timing of individual unforced climate events like El Nino years in the historical runs will rarely (and only by chance) coincide with years of actual occurrence, since historical runs are initiated from an arbitrary point of quasi-equilibrium control run. Hence, the results should be analyzed in probabilistic terms in a similar manner as [18–20].

GCMs that were used for both RCP8.5 and RCP4.5 simulations were applied to project the precipitation patterns over Lake Victoria for the 2040s (2020–2060) and 2075s (2055–2095). The historical and future precipitation for the 2040s and 2075s was obtained by simply averaging the simulations from the different GCMs. This method is opposed to applying weighting factors described in [21] and was applied to avoid introducing extra uncertainties as tested by [22]. A total of 42 GCM runs obtained from 26 models simulated by 16 different modeling centers of the CMIP5 archive were used. GCM simulations for the historical period were obtained for the different quarters of Lake Victoria (Figure 1) and averaged to obtain the areal precipitation over the lake.

The performance of GCMs was evaluated based on the historical outputs using the absolute observed precipitation series over the lake for the 41-year period 1962–2002 provided by [12]. Precipitation measurements over the lake are sparse and of low quality. Kizza [16] compared satellite measurements to the lake surrounding observations—TRMM 3B43 product improved the quality of precipitation over the lake by 33% while the PERSIANN product improved the precipitation series by 76%. Kizza *et al.* [12] improved the spatial precipitation input using gridded monthly precipitation with a spatial resolution of 2 km for both ground based and satellite data for the period 1960–2004 providing a plausible lake balance model (Figure 2).

Due to the variation in GCM outputs and for clearer analysis of results, the precipitation simulations by the GCMs were further classified according to the GCM grid sizes. The spatial resolution of the CMIP5 coupled models range from 0.5° to 4° for the atmospheric component and 0.2° to 2° for the ocean component [17]. Table 1 shows an overview of the 26 GCMs used in this study. The model resolutions are classified as follows: Low Resolution (LR) models: grid size >2°, Medium Resolution (MR) models: 1.5° to 2°, High Resolution (HR) models: <1.5° based on their seasonal performance.

Table 1. Coupled Model Intercomparison Project (CMIP5) general circulation models (GCMs) considered in this study; blue (b), green (g) and red (r).

	Modeling Center	Country	Model	Lat.	Lon.	Res.	Color
i.	Commonwealth Scientific and Industrial Research Organization/ Bureau of Meteorology (CSIRO-BOM)	Australia	ACCESS1.0	1.87	1.25	MR	g
ii.	College of Global Change and Earth System Science, Beijing Normal University	China	BNU-ESM	2.81	2.79	LR	r
iii.	*Centro Euro-Mediterraneo per I Cambiamenti Climatici*	Italy	CMCC-CESM	3.75	3.71	LR	r
		Italy	CMCC-CMS	1.87	1.87	MR	g
iv.	*Centre National de Recherches Meteorologiques / Centre Europeen de Recherche et Formation Avancees en Calcul Scientifique (CNRM/CERFACS)*	France	CNRM-CM5	1.41	1.40	HR	b
v.	Commonwealth Scientific and Industrial Research Organization/ Queensland Climate Change Centre of Excellence (CSIRO-QCCCE)	Australia	CSIRO-Mk3.6	1.87	1.87	MR	g
vi.	Canadian Centre for Climate Modelling and Analysis	Canada	CanESM2	2.81	2.79	LR	r
vii.	Geophysical Fluid Dynamics Laboratory	US-NJ	GFDL-ESM2G	2.5	2.0	LR	r
		US-NJ	GFDL-ESM2M	2.5	2.0	LR	r
viii.	NASA Goddard Institute for Space Studies	US-NY	GISS-E2-H	2.5	2.0	LR	r
		US-NY	GISS-E2-R	2.5	2.0	LR	r
ix.	Met Office Hadley Centre	UK-Exeter	HadCM3	3.75	2.5	LR	r
		UK-Exeter	HadGEM2-CC	1.87	1.25	MR	g
		UK-Exeter	HadGEM2-ES	1.75	1.25	MR	g
x.	*Institut Pierre-Simon Laplace*	France	IPSL-CM5A-LR	3.75	1.89	LR	r
		France	IPSL-CM5A-MR	2.50	1.26	LR	r
		France	IPSL-CM5B-LR	3.75	1.89	LR	r

Table 1. *Cont.*

	Modeling Center	Country	Model	Lat.	Lon.	Res.	Color
xi.	Atmosphere and Ocean Research Institute (The University of Tokyo), National Institute for Environmental Studies, and Japan Agency for Marine-Earth Science and Technology	Japan	MIROC-ESM	2.81	2.79	LR	r
		Japan	MIROC5	1.40	1.40	HR	b
xii.	Max Planck Institute for Meteorology (MPI-M)	Germany	MPI-ESM-LR	1.87	1.87	MR	g
		Germany	MPI-ESM-MR	1.87	1.87	MR	g
xiii.	Meteorological Research Institute	Japan	MRI-CGCM3	1.12	1.12	HR	b
xiv.	Norwegian Climate Centre (NCC)	Norway	NorESM1-M	2.50	1.89	LR	r
xv.	Beijing Climate Center, China Meteorological Administration	China	BCC-CSM1.1m	1.12	1.12	HR	b
		China	BCC-CSM1.1	2.81	2.79	LR	r
xvi.	Institute for Numerical Mathematics	Russia	INM-CM4	2.0	1.5	MR	g

2.4. Model Performance Evaluation

Probabilistic analyses were performed to evaluate the effect of climate change on absolute precipitation for different aggregation scales *i.e.*, yearly, monthly, daily; and to investigate reasons for the precipitation changes. The GCM performance was analyzed for the different seasons *i.e.*, January-February (JF), March-May (MAM), June-September (JJAS) and October-December (OND) based on the Normalized Mean Error (NME) and the covariance between the observed and historical GCM output. The NME is defined as the ratio of mean error to sample mean of the observations, while covariance is a measure of how two variables change together—positive covariance implies variables increase or decrease together. Evaluation of GCM performance for annual precipitation was based on the Coefficient of Variation of the Root Mean Square Error (CV(RMSE)) as well. The CV(RMSE) was computed as the ratio of the RMSE to the mean of the observations. The ability of the GCMs to simulate high and extreme precipitation was checked for daily, monthly and annual time scales. For that purpose, precipitation amounts were ranked and plotted against the empirical return period to determine how well the GCMs perform in extreme precipitation distributions. This analysis is useful from a water engineering point of view: If the GCM results would be used for obtaining water engineering design or planning values in terms of precipitation amount for given return periods, the analysis shows the deviations that can be found in these design or planning values.

One important remark should be made about this GCM performance evaluation based on historical precipitation observations: model performance for the historical period, as evaluated here, is not equivalent to future model performance. The latter obviously cannot be validated; that is why the historical analysis is used instead as indicative for future performance.

To determine the influence of future climate change on precipitation, the ratios of potential future simulated precipitation to historical precipitation simulations—hereafter referred to as perturbation factors, were used to project impacts of climate change in the Lake Victoria basin. This approach has been applied to study climate change by several authors e.g., [19,23]. The source of future changes in precipitation reflected in the perturbation factors was further analyzed in the different seasons to understand the influence of individual effects like changes in intensities or number of wet days in each season on the global annual change using Box plots. This analysis aims to provide plausible quantifiable measures of precipitation changes over Lake Victoria in the 21st century.

3. Results and Discussion

3.1. GCM Performance Evaluation

3.1.1. Monthly, Seasonal and Annual Precipitation

The GCM historical and observed series for the period 1962–2002 were aggregated over monthly time scales to evaluate the seasonal variations in the model based precipitation amounts and how much they deviate from the absolute observed values. This was done for the different resolution GCMs (Figure 4a). LR GCMs with grid sizes greater than 2° (>220 km) generally fail to simulate the wetter MAM rainy season depicted by the observed series while the HR GCMs (<165 km) show an acceptable seasonal pattern (Figure 4a). Based on the precipitation results only, the performance of the GCMs improved with the increase in resolution of the GCMs (Figure 4a). The different runs within the same GCMs did not necessarily produce a discrepancy as large as that between the different GCMs (Figure 4a) implying that model parameterization is probably more vital in determining GCM output compared to GCM initializations. For example, CanESM2.1, CanESM2.2, CanESM2.3 have different initializations but not different parameters compared to another model e.g., HadGEM2-CC.1 and HadGEM2-CC.2. From Figure 4a, we can see that differences between HadGEM-CC and CanESM2 models are larger than those arising between different runs within the same model.

Earlier research by [19] reported that there was no strong evidence to suggest that GCM performance improved with higher spatial resolution for the previous generation GCMs (4th Assessment Report of the IPCC based on CMIP3). Of the 18 GCMs of CMIP3 used by [19], only CCSM3.0 can be categorized as HR, based on the definition used in the present manuscript. The three other CMIP3 GCMs (MK3.0, MK3.5, and ECHAM5) similar to CSIRO-Mk3.6 and MPI-ESM-LR in CMIP5 are categorized as MR while the rest fall under LR models. HR and MR GCMs such as CCSM3.0, MK3.0, MK3.5 and ECHAM5 were ranked in the top five performing GCMs while most other GCMs performed poorly for the Katonga and Ruizi catchments, which are located within the Lake Victoria basin [19]. This study conforms to our hypothesis even though the areal extent of these catchments was in the order of 1000–3000 km^2 [19] compared to the 68,800 km^2 expanse of the lake. The improvement in the CMIP5 simulations in which higher spatial resolution coupled models were used to obtain a richer set of outputs cannot be neglected—however, IPCC [8] recognizes an undisputed similarity between CMIP3 and CMIP5 model simulations. This implies that model resolution was vital in determining the GCM performance.

Underestimation of monthly precipitation totals for the LR GCMs can be attributed to the large grid sizes that do not allow simulating different precipitation patterns over the northern and southern parts of the lake since rainfall patterns vary across the Lake Victoria basin. The universal kriging and inverse distance weighting methods used by [16] to obtain the spatial distribution of precipitation over the Lake Victoria basin show influence of the seasonal migration of the ITCZ on the rainy seasons such that the north eastern region generally receives more precipitation compared to the south eastern region. The GCMs underestimate precipitation in the rainy MAM season and the dry JJAS season, but overestimate the variable OND rainy season (Figure 4a,b). With the exception of the HadCM3 model, most GCMs simulate well the variable OND rainy season, which is highly influenced by complex interactions between the Indian and Pacific Oceans, a phenomenon that is well captured by the GCMs. HadCM3, HadGEM2-CC and HadGEM2-ES are developed by the same modeling center using similar radiative forcing. Although these models are essentially different, improved seasonal patterns are noticed in the finer HadGEM2-CC and HadGEM2-ES models compared to the coarser HadCM3 model (Figure 4a).

Figure 4. (**a**) Average monthly precipitation for the different GCMs compared to the observed series, red: Low resolution (LR) GCMs; blue: High resolution (HR) GCMs; green: Medium resolution (MR) GCMs; (**b**) Difference between modeled and observed precipitation for the different classifications of GCMs.

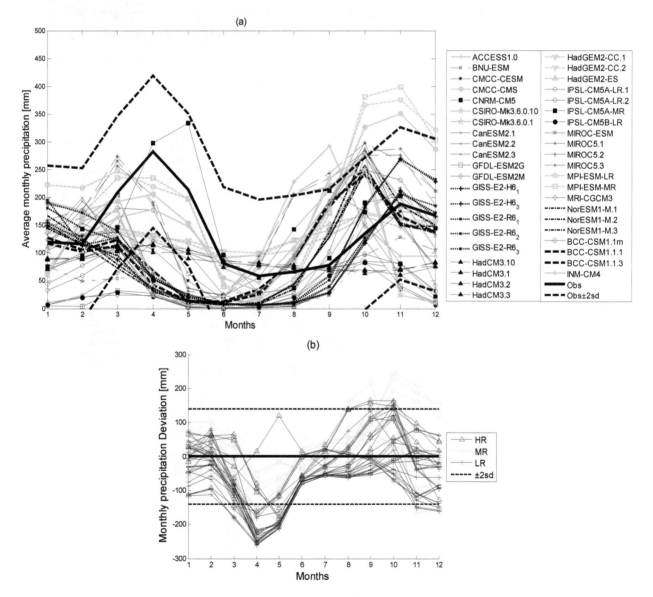

There is a one month lag in the rainy seasons simulated by the GCMs as compared to the observed precipitation (Figure 4a). The monthly precipitation anomalies were calculated to account for the climatology simulated by the different GCMs as a difference between average monthly simulated precipitation, and the average monthly observed precipitation (Figure 4b). Most GCMs simulate the June-February precipitation well (lower monthly anomaly values) while the LR GCMs generally underestimate the MAM season even though there are more LR models compared to HR and MR models (Figure 4b). A more general seasonal check was applied in the JF, MAM, JJAS and OND seasons—to even out effects of the time lag exposed in Figure 4a, as shown in Figure 5a. The LR GCMs generally show lower (often negative) NME and covariance closer to zero or more negative implying that the observed and simulated historical seasonal precipitations did not change together for the LR models. From the covariance results (Figure 5b), it can be concluded that on average the tendency of a linear relationship between the observed and simulated seasonal precipitation decreased with increase in the grid size of the model. Most LR models showed negative covariance, while most HR models showed positive covariance for the average seasonal precipitation.

Figure 5. (**a**) Average seasonal precipitation for the different GCMs compared to the observed series for January-February (JF), March-May (MAM), June-September (JJAS) and October-December (OND) aggregations; (**b**) Covariance *vs*. normalized mean error (NME) for seasonal averages, red: LR GCMs; blue: HR GCMs; green: MR GCMs.

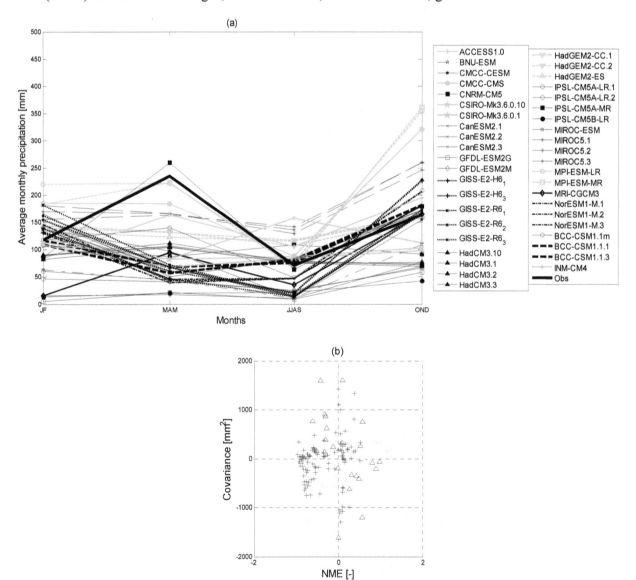

Figure 6 shows the NME and its statistical significance compared to the uncertainty bounds approximated by twice the normalized standard deviation to approximate a 95% confidence interval. The best performing GCMs are again the higher resolution GCMs: CNRM-CM5, ACCESS1.0 and MRI-CGCM3, which lay within plotted uncertainty bounds. Figure 6 also shows the CV(RMSE) on the annual precipitation amounts. The GCMs: GFDL-ESM2G, GFDL-ESM2M, IPSL-CM5A-LR and IPSL-CM5B-LR (all of which are LR models) produce the highest CV(RMSE) and NME, henceforth are considered to be poorly performing. Generally, the GCMs perform better with the annual precipitation simulations compared to seasonal and monthly aggregations based on the NME (−1 to 2.5 for monthly; and −0.7 to 0.5 for annual aggregations in Figure 6).

Figure 6. NME and coefficient of variation of the root mean square error (CV(RMSE)) of average annual precipitation for the GCM simulations compared to the observed series, red: LR GCMs; blue: HR GCMs; green: MR GCMs.

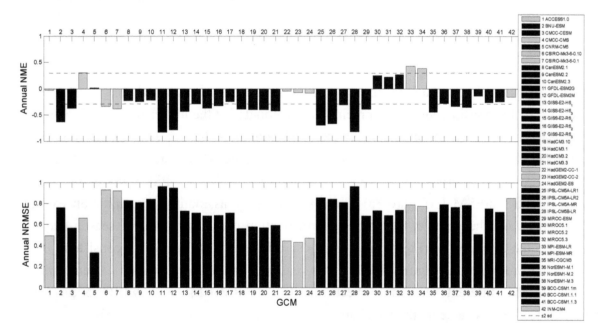

The GCMs show acceptable annual precipitation patterns (observed values located within the interval defined by the standard deviation of the GCM ensemble) but fail to simulate the peak precipitation (Figure 7a,b). The peak precipitation seasons are not well simulated in the GCMs; due to the inability of the GCMs to capture the heavy MAM precipitation even though model representation improves with increased model resolution. Although the peak annual precipitation is not well captured, the general annual variability trend is typically reproduced as it depends on the well simulated OND rainy season rather than the heavy MAM season. The OND and MAM seasons account for more than 65% of the total annual precipitation over the lake, however the variability of precipitation in the OND period has a greater influence on the annual precipitation compared to that in the MAM period [16]. The correlation coefficients between seasonal and annual precipitation totals for the OND and MAM periods were 0.71 and 0.5 respectively, *i.e.*, peaks in annual precipitation totals tended to coincide with peaks in OND rather than MAM seasonal precipitation [24].

Figure 7. Annual variation of observed, and GCM output for (**a**) RCP4.5; (**b**) RCP8.5, GCM output bounds are based on twice the standard deviation.

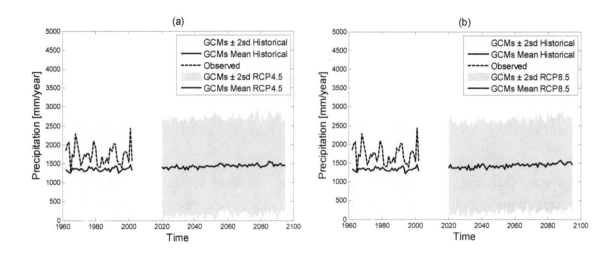

3.1.2. Precipitation Extremes

The LR GCMs underestimate the monthly and annual precipitation amount for given return periods (Figure 8). The underestimation of the precipitation amounts for the higher return periods is probably due to the large variations in topographical and areal properties that are evened out over wider areas while HR GCMs generally provide better simulations for monthly extremes. Some LR GCMs provide satisfactory monthly precipitation extremes, notably BCC-CSM1.1, IPSL-CM5A-MR, GISS-E2-H GISS-E2-R and NorESM-1 (Figure 8a)—this is misleading as monthly precipitation extremes are selected throughout the year yet the observed precipitation peaks in the MAM season may coincide with those in the OND season (Figure 4a). It is not surprising that the LR GCMs consistently simulated lower annual precipitation (Figure 8b). For this reason, monthly precipitation extremes are further classified in the different seasons in Section 3.1.2. CNRM-CM5 and MIROC5 gave the best estimations for both extreme monthly and annual precipitation as shown in Figure 8b.

The GCM performance was evaluated in the wet and dry months *i.e.*, November and July respectively (Figure 9). For the rainy November, the uncertainty in simulating daily precipitation extremes with return periods higher than 4 years is very large, irrespective of the model resolution. The HR models overestimate the extreme precipitation amounts in the wet month of November. Many of these extreme events in the observed series are related to occurrence of El-Niño years as precipitation in the region is strongly quasi-periodic with a dominant ENSO timescale of variability of 5–6 years [9]. The monthly shift in the seasonal variations for the BCC-CSM1.1m, MRI-CGCM3 and BCC-CSM1.1 models was depicted in the daily extreme plots (Figure 9). These HR models overestimate extreme daily precipitation; yet in reality it is due to the one month time lag (Figure 4a). However, in July (the driest month), the GCM performance is very erratic with most LR GCMs underestimating the daily extremes. The uncertainty is higher in the driest month of July which experiences largely varying precipitation (Figure 9b).

Figure 8. (**a**) Return period of average monthly precipitation; (**b**) Average annual precipitation for the different GCM simulations as compared to the observed series; red: LR GCMs; blue: HR GCMs; green: MR GCMs.

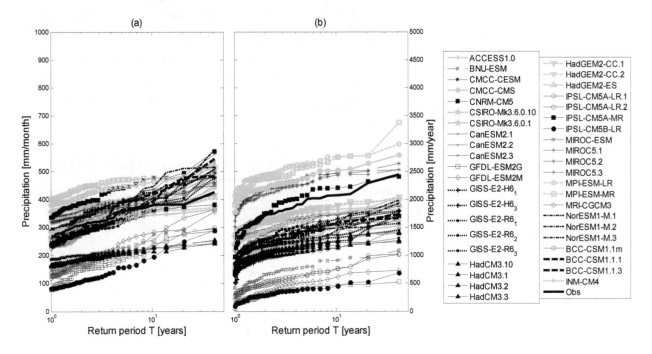

Figure 9. Return period of average daily precipitation for the different GCM simulations for November and July, red: LR GCMs; blue: HR GCMs; green: MR GCMs.

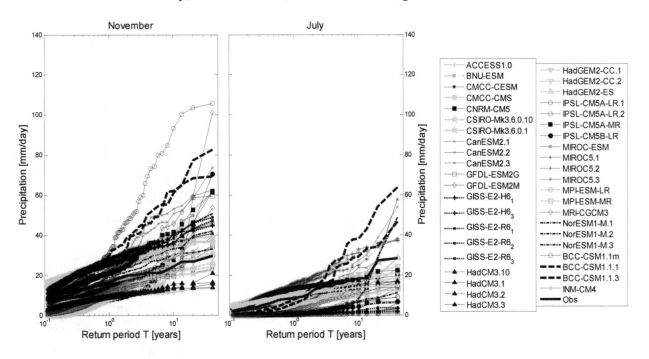

The GCM evaluation showed that GCM outputs provide better results in terms of annual and seasonal precipitation compared to the daily scale analyses (Figures 8,9). Even if model performance cannot be evaluated based on a single index, an array of measures, such as those described in this section provide a good indication of the model overall performance. Higher resolution models provide better estimates of annual, seasonal and monthly precipitation; and precipitation variation (Figures 4,5,8). Table 2 provides a ranking of the good and poor performing GCMs based on annual, seasonal and monthly performance. Differences in latitudes are more significant in GCM precipitation performance than the differences in longitudes. CNRM-CM5 provides the best estimate for annual precipitation and seasonal variation with a minimum time shift in monthly simulations while HadCM3 fails to describe even the basic seasonal variation. As shown by Shaffrey *et al.* [25], reduction of the horizontal resolution e.g., in the HadGEM1 model, may result in reduced SST errors and more realistic approximations of small scale processes, especially the ENSO phenomenon leading to improvement of results simulated by the finer HiGEM model.

Table 2. Ranking of CMIP5 GCMs based on simulation of precipitation over Lake Victoria.

Good Performing GCMs	Long.	Lat.	Resolution	Poor Performing GCMs	Long.	Lat.	Resolution
CNRM-CM5	1.41	1.40	High	HadCM3	3.75	2.5	Low
MIROC5	1.40	1.40	High	IPSL-CM5A-LR	3.75	1.89	Low
BCC-CSM1.1m	1.12	1.12	High	IPSL-CM5B-LR	3.75	1.89	Low
ACCESS1.0	1.87	1.25	Medium	GFDL-ESM2G	2.5	2.0	Low
HadGEM2-CC	1.87	1.25	Medium	GFDL-ESM2M	2.5	2.0	Low
HadGEM2-ES	1.75	1.25	Medium				

3.2. Analysis of Projected Future Precipitation by GCMs

3.2.1. Monthly, Seasonal and Annual Precipitation

The analysis of projected future changes in rainfall shows no significant change for the average monthly precipitation in the 2040s, but a slight increase for the 2075s especially towards the end of the shorter OND rainy season (Figure 10). The historical analysis described earlier in Section 3.1 showed that the precipitation in the OND season is well captured by the GCMs. The magnitude of change is slightly higher for RCP8.5 under which the temperature increase is higher, leading to higher evapotranspiration and precipitation. The effect of GCM uncertainty is found to be far greater than that due to precipitation simulations between the RCP8.5 and RCP4.5 scenarios (Figure 10). Uncertainties in the future simulations are higher for the 2075s than for the 2040s as scenario uncertainty attributed to the uncertainty in emissions of greenhouse gases—hence radiative forcing increases exponentially especially after the 2060s [26].

Perturbation factors for annual precipitation due to climate change are shown in Figure 11. Generally annual precipitation changes converge to the same level for precipitation of return periods greater than two years. For that reason and to simplify the presentation of results, the mean change is computed for the precipitation extremes and plotted for all GCMs in box plots (Figure 11a). Precipitation extremes are defined as events that are larger or equal to those that occur at least once a year. The lower resolution GCMs like IPSL-CM5A-LR, IPSL-CM5B-LR, BNU-ESM, GFDL-ESM2G and GFDL-ESM2M show higher precipitation changes and mostly positive, while the higher resolution GCMs like CNRM-CM5, BCC-CSM1.1m and MIROC5 show precipitation crowding around the unchanged mean climate.

Figure 10. Seasonal variation of average monthly precipitation for observed and GCM historical, RCP4.5 and RCP8.5 series for (**a**) 2040s; (**b**) 2075s. The colored dotted lines indicate the extent of twice the historical standard deviation.

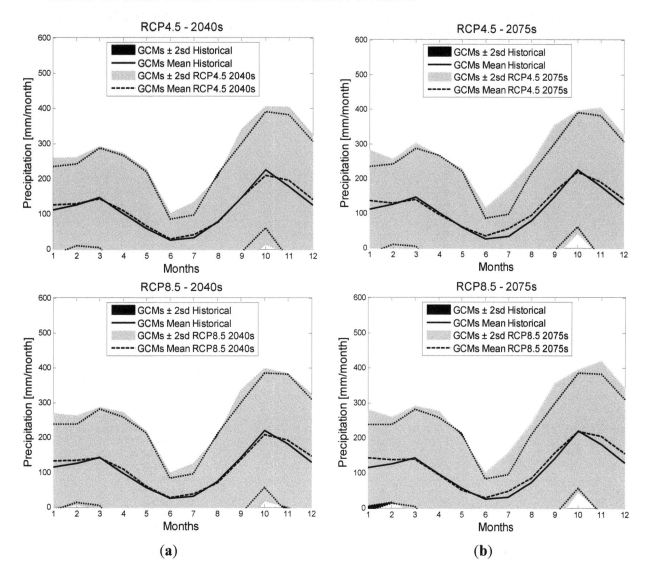

(**a**)   (**b**)

Figure 11 shows an increase in annual precipitation over Lake Victoria in the 21st century for both RCP4.5 and RCP8.5 scenarios (only RCP8.5 shown). For the 2040s, annual precipitation is projected to increase by about 7% for both scenarios, while for the 2075s it is expected to increase by about 10% for the RCP4.5 scenario, and more than 15% for the RCP8.5 scenario. Next to this analysis of annual precipitation, summations of precipitation in the MAM, OND and JJAS seasons were analyzed to determine the perturbation factors for the 2040s and 2075s in order to understand the effect of seasonal precipitation on annual precipitation over the Lake Victoria basin. Figure 12 shows the seasonal change factors for RCP8.5 scenario. Most GCMs generally agree well in the OND rainy season as depicted by the lower divergence and narrower box limits for all resolutions (Figure 12a). Precipitation amounts generally increase in all the seasons for the mitigation RCP4.5 scenario. However, for the RCP8.5 scenario, the seasonal amounts increase only in the rainy seasons (Figure12a). For the dry JJAS season in RCP8.5 scenario, the total seasonal precipitation amount is expected to decrease by about 10% in the 2040s, and increase by about 20% in the 2075s.

Figure 11. (**a**) Perturbation factors for annual precipitation using events with return periods greater than two years for the different GCMs, for the 2040s (notched) and 2075s for RCP8.5; (**b**) Perturbation factors *vs.* return period for annual precipitation for the different GCM simulations for the 2075s under RCP8.5 scenario: Red = LR GCMs, Blue = HR GCMs and Green = MR GCMs.

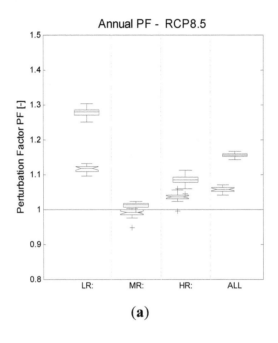

(**a**)

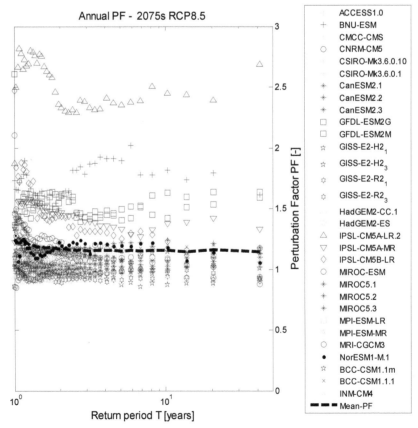

(**b**)

Figure 12. (**a**) Perturbation factors for total seasonal precipitation simulated by the different GCMs for the 2040s (notched) and 2075s, for RCP8.5; (**b**) Perturbation factors *vs.* return period for total seasonal precipitation simulated by the different GCMs in the 2075s for RCP8.5, Red = LR GCMs, Blue = HR GCMs and Green = MR GCMs.

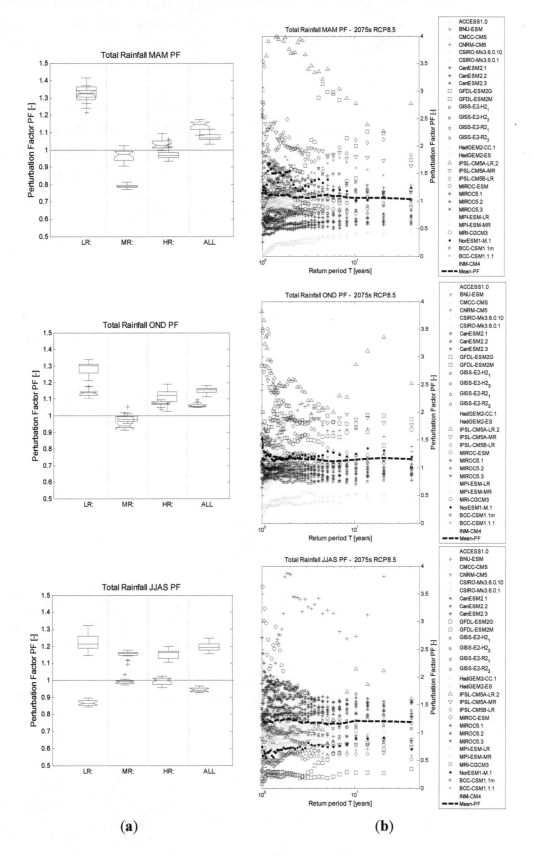

(**a**)   (**b**)

3.2.2. Precipitation Extremes

Changes in Number of Wet Days

A wet day is defined as that having intensity greater than 0.1 mm/day. Precipitation volumes are affected by both the number of wet days and the intensity of precipitation. The number of wet days in the historical and future scenarios was obtained for the different seasons to determine the relative changes in the wet day frequency (Figure 13).

Figure 13. (a) Perturbation factors for number of wet days simulated by GCMs in the different seasons for the 2040s (notched) and 2075s, for RCP8.5; **(b)** Perturbation factors *vs.* return period for number of wet days simulated by the GCMs in the 2075s for RCP8.5 in different seasons, Red = LR GCMs, Blue = HR GCMs and Green = MR GCMs.

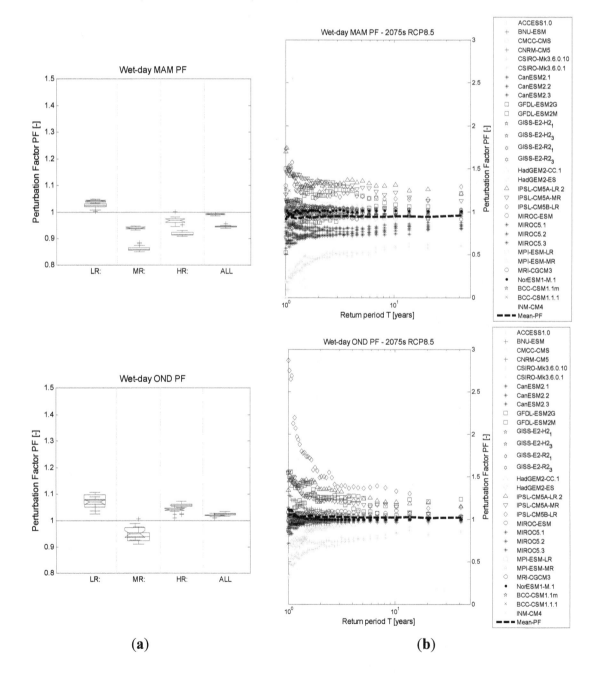

(a)   (b)

Figure 13. *Cont.*

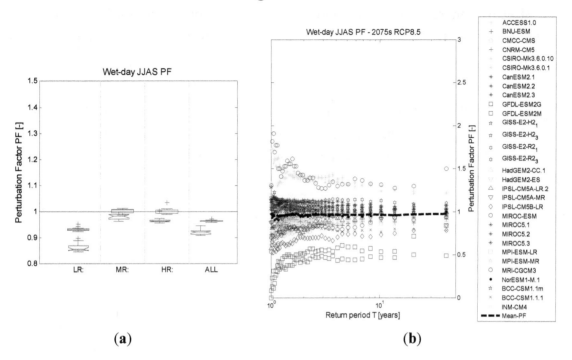

(a)   (b)

   The number of wet days decreases in the MAM and JJAS seasons but slightly increases in the OND rainy season for both RCP4.5 and RCP8.5 scenarios (Figure 13). The LR GCMs project greater changes in the number of wet days for the rainy MAM and OND seasons but simulate lower number of wet days in the dry JJAS season probably because larger areas even out localized low precipitation intensities. Annual precipitation over Lake Victoria is estimated to be about 26% higher than over land. This is expected to be associated with 20%–30% more occurrences of cold cloud tops over the lake [27]. It implies that averaging over large areas including land is bound to reduce the precipitation over the lake and sometimes the number of wet days especially in the dry seasons. This again confirms that GCM parameterization and resolution have an important effect on GCM outputs.

Changes in Wet Day Intensities

   The daily precipitation intensities in the MAM, OND and JJAS seasons were analyzed to determine the change factors in the 2040s and 2075s under the different scenarios. Figure 14a shows the changes in daily precipitation intensities for the different seasons for events with return periods greater than two years. When changes in wet day intensities *vs.* return periods are analyzed, the intensities are generally seen to increase in the rainy seasons for both RCP4.5 and RCP8.5 scenarios. The daily precipitation extremes increase more towards the 2075s compared to the 2040s especially for RCP8.5 (Figure 14). However, for the dry JJAS season in the RCP8.5 scenario, daily precipitation is generally expected to remain constant in the 2040s and increase by more than 20% in the 2075s explaining the reason for the decrease in the JJAS seasonal precipitation in the 2040s, and increase in the 2075s (Figure 12). The number of wet days slightly decreases in the dry seasons for both decades yet daily precipitation intensities seem to increase more in the late century than the mid-century. In the 2040s, the RCP8.5 and RCP4.5 projections anticipate similar changes in daily extreme precipitation. The difference in relative forcing for the two scenarios in the 2040s is 1 Wm^{-2} compared to 3 Wm^{-2} in the 2075s

(Figure 3). This relative difference is consistent with higher temperatures in the 2075s that encourage formation of heavier intense convective storms in the dry season as more moisture is stored in the atmosphere. Therefore, RCP8.5 suggests fewer but heavier intense storms in the 2075s if carbon emissions are not controlled.

Figure 14. (**a**) Perturbation factors for daily precipitation simulated by GCMs in different seasons for the 2040s (notched) and 2075s, for RCP8.5; (**b**) Perturbation factors *vs.* return period for daily precipitation simulated by the GCMs in the 2075s for RCP8.5 in the different seasons, Red = LR GCMs, Blue = HR GCMs and Green = MR GCMs.

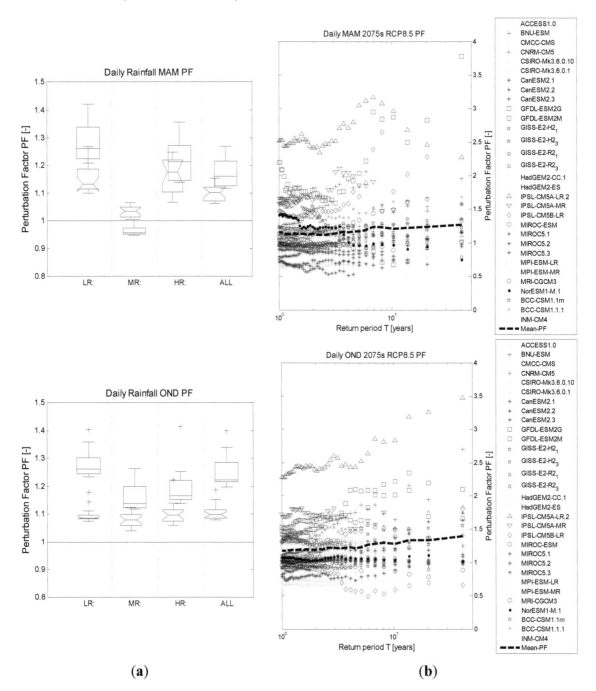

(**a**)   (**b**)

Figure 14. *Cont.*

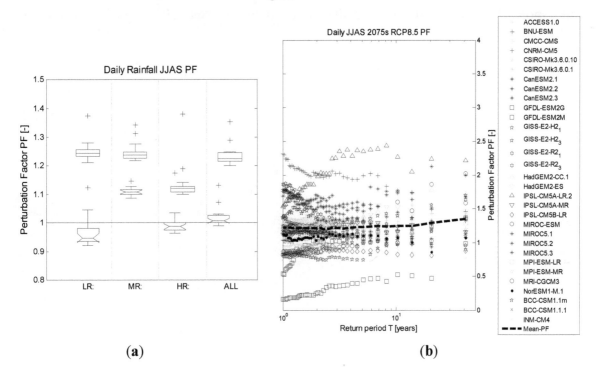

(a)   (b)

Most models show an increase in daily precipitation intensities for the dry JJAS season with some LR GCMs like GFDL-ESM2G, GFDL-ESM2M and IPSL-CM5A-LR strongly deviating from the mean change (Figure 14b). The difference in GCM performance are larger in the JJAS and MAM seasons—which were not well captured by the GCMs. Daily precipitation intensities are expected to increase by about 10%–25% in the OND season, which is consistent with the 10%–20% increase in total precipitation for that season (Figure 12). Despite the 10%–15% increase in MAM daily extremes, seasonal volumes increase by less than 10% in the same season since the number of wet days generally decreases in this season (Figure 13). Notwithstanding, in reality this increment is not expected to have any significant influence on the annual precipitation volumes especially since precipitation in the MAM season is generally underestimated by the GCMs as explained in Section 3.1.

Daily precipitation intensities were also checked in the wettest months (April and November) and dry July month (Figure 15). The large spread of the perturbation factor quartiles in the dry month of July is attributed to division by very low historical rainfall amounts especially for LR GCMs that provide precipitation results averaged over larger areas (Figure 15). The resolutions of the GCMs affect the GCM output so care ought to be taken when choosing GCMs for climate change impact projections. LR GCMs show very large variations from the mean change while the HR GCMs values are crowded around the mean. The large variations are even more pronounced for the RCP8.5 scenario. The large uncertainty in the GCM output for LR models is carried into the computed value for the mean change especially visible in the dry month of July (Figure 15), even when mean change often coincides with the results from the HR GCMs suggesting that finer resolution GCMs are favorable in predicting climate change scenarios.

Figure 15. Perturbation factors for daily precipitation simulated by GCMs in the months of April, November and July for the 2040s (notched) and 2075s, including outliers represented by (+) (**a**) RCP4.5; (**b**) RCP8.5.

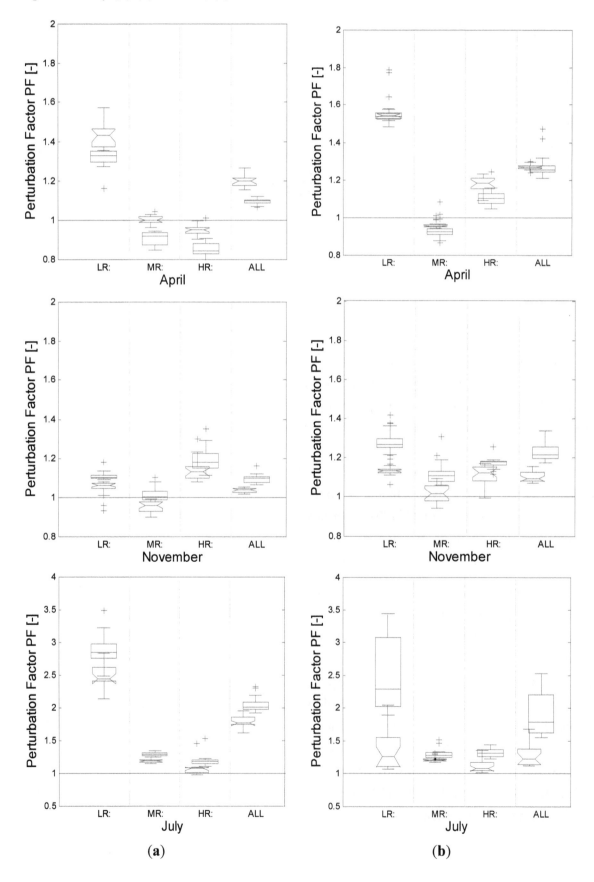

3.3. RCP4.5 vs. RCP8.5 and 2040s vs. 2075s Comparison

Figures 16 and 17 summarize the differences in GCM results for the different scenarios and periods based on their resolutions. Precipitation will generally increase in the 21st century for both RCP4.5 and RCP8.5 scenarios—higher increase is generally anticipated for RCP8.5 compared to RCP4.5 (Figure 16). For RCP4.5, annual precipitation is expected to increase by about 7% for both 2040s and 2075s, while RCP8.5 projects about 10% increase in the 2040s and about 18% in the 2075s. This increase is generally attributed to increased precipitation intensities rather than the total number of wet days, as heavier intense storms are expected in the late 21st century according to Section 3.2. The results are consistent with the positive shift in precipitation distribution expected in other parts of East Africa under global warming for the CMIP3 climate models [28]. Generally, the increase in precipitation is more for the 2075s than for 2040s; and this effect is even greater than that arising from differences between RCP8.5 and RCP4.5 scenarios (Figure 17). A high level of uncertainty is presented by the LR GCMs (grid size >2°) compared to the HR and MR GCMs (Figure 17a,b).

Figure 16. Comparison of mean perturbation factors for annual precipitation based on GCM resolutions for the 2040s (notched) and 2075s, (**a**) RCP4.5; (**b**) RCP8.5.

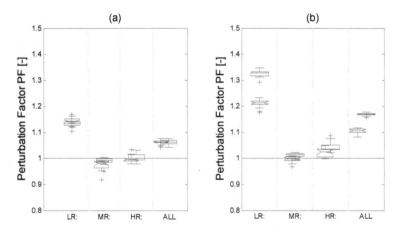

Figure 17. Annual precipitation perturbation factors, (**a**) RCP8.5 *vs.* RCP4.5, for the 2040s (o) and 2075s (+); (**b**) 2075s *vs.* 2040s for RCP4.5 (Δ) and RCP8.5(x), Red = LR GCMs, Blue = HR GCMs and Green = MR GCMs.

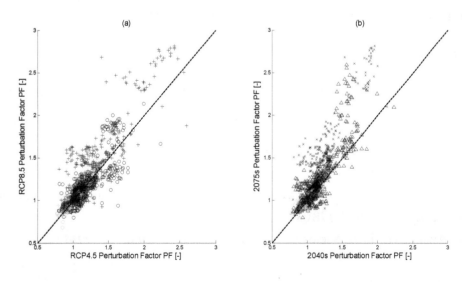

4. Conclusions

The GCM performance over Lake Victoria is highly dependent on the resolution of the GCM, especially the latitudinal scale. High resolution GCMs, namely CNRM-CM5, MIROC5, HadGEM2-CC, HadGEM2-ES and BCC-CSM1.1m gave the best performance in modeling past absolute precipitation over Lake Victoria. Lower resolution GCMs (grid size >2.5°) e.g., GFDL-ESM2G, GFDL-ESM2M, IPSL-CM5A-LR, IPSL-CM5B-LR and HadCM3 produced larger uncertainties in precipitation simulations. Therefore, for future projections of precipitation, high resolution GCMs are favored to provide reliable seasonal results. However, there is need to use a wide range of GCMs, irrespective of their resolution in order to sufficiently capture the uncertainty in climate modeling physics. This uncertainty may be large as also shown in this paper; hence needs to be taken into account in hydrological impact investigations of climate change.

The total annual precipitation is expected to increase by about 6%–8% for the RCP4.5 scenario and about 10%–18% for the RCP8.5 scenario over the 21st century, despite the higher (up to 40%) increase in extreme daily intensities since the number of wet days does not significantly change. This increase is expected to be higher in the late 21st century (2075s) than in the 2040s.

To study future lake level changes, next to precipitation over the lake, also discharges of main inflowing rivers need to be studied. This requires future projections of precipitation and potential evapotranspiration over the Lake Victoria Basin river subcatchments, and impact modeling by means of catchment runoff models. This research provided the baseline for such study by conducting GCM evaluations in precipitation simulation and analysis of future projections.

Acknowledgments

The authors would like to thank the *Vlaamse Interuniversitaire Raad*—Institutional University Cooperation (VLIR-IUC) of Belgium for funding this research under the VLIR-ICP PhD program. We also would like to thank the Ministry of Water and Environment, Uganda, and CMIP5 project for availing the data for this research. The performance evaluations in this study are based on the daily CMIP5 model output, obtained through the Program for Climate Model Diagnosis Intercomparison (PCMDI) portal of the Earth System Grid Federation [29]. We also acknowledge use of the precipitation series dataset provided by Michael Kizza. We thank the anonymous reviewers for their constructive comments that tremendously improved the quality of this paper.

Author Contributions

Patrick Willems and Charles B. Niwagaba supervised the PhD research and reviewed this paper contributing to the write-up.

References

1.   Kite, G.W. Recent changes in level of Lake Victoria/Récents changements enregistrés dans le niveau du Lac. *Hydrol. Sci. J.* **1981**, *26*, 233–243.
2.   Kite, G.W. Analysis of Lake Victoria levels. *Hydrol. Sci. J.* **1982**, *27*, 99–110.

3. Piper, B.S.; Plinston, D.T.; Sutcliffe, J.V. The water balance of Lake Victoria. *Hydrol. Sci. J.* **1986**, *31*, 25–37.

4. Yin, X.; Nicholson, S.E. The water balance of Lake Victoria. *Hydrol. Sci. J.* **1998**, *43*, 789–811.

5. Tate, E.; Sutcliffe, J.; Conway, D.; Farquharson, F. Water balance of Lake Victoria: Update to 2000 and climate change modelling to 2100/Bilan hydrologique du Lac Victoria: Mise à jour jusqu'en 2000 et modélisation des impacts du changement climatique jusqu'en 2100. *Hydrol. Sci. J.* **2004**, *49*, doi:10.1623/hysj.49.4.563.54422.

6. Awange, J.L.; Ogalo, L.; Bae, K.-H.; Were, P.; Omondi, P.; Omute, P.; Omullo, M. Falling Lake Victoria water levels: Is climate a contributing factor? *Clim. Change* **2008**, *89*, 281–297.

7. Swenson, S.; Wahr, J. Monitoring the water balance of Lake Victoria, East Africa, from space. *J. Hydrol.* **2009**, *370*, 163–176.

8. Intergovernmental Panel on Climate Change (IPCC). *CLIMATE CHANGE 2013: The Physical Science Basis. Contribution of Working Group I to the Fifth Assessment Report of the Intergovernmental Panel on Climate Change*; Cambridge University Press: Cambridge, United Kingdom and New York, NY, USA, 2013; p. 1535.

9. Nicholson, S.E. A review of climate dynamics and climate variability in Eastern Africa. In *Limnology, Climatology and Paleoclimatology of the East African Lakes*; Johnson, T.C., Odada, E.O., Eds.; Gordon and Breach Publishers: Amsterdam, The Netherlands, 1996; pp. 25–56.

10. Indeje, M.; Semazzi, F.H.M.; Ogallo, L.J. ENSO signals in East African rainfall seasons. *Int. J. Climatol.* **2000**, *20*, 19–46.

11. Nicholson, S.E.; Yin, X. Rainfall conditions in Equatorial East Africa during the Nineteeenth century as inferred from the record of Lake Victoria. *Clim. Chang.* **2001**, *48*, 387–398.

12. Kizza, M.; Rodhe, A.; Xu, C.-Y.; Ntale, H.K.; Halldin, S. Temporal rainfall variability in the Lake Victoria Basin in East Africa during the twentieth century. *Theor. Appl. Climatol.* **2009**, *98*, 119–135.

13. Sene, K.J.; Plinston, D.T. A review and update of the hydrology of Lake Victoria in East Africa. *Hydrol. Sci. J.* **1994**, *39*, 47–63.

14. Mutenyo, I.B. Impacts of Irrigation and Hydroelectric Power Developments on the Victoria Nile in Uganda School. Ph.D. Thesis, Cranfield University, Cranfield, UK, 2009; p. 258.

15. Sutcliffe, J.; Parks, Y. *The Hydrology of the Nile*; International Association of Hydrological Sciences IAHS: Wallingford, UK, 1999; Volume 5.

16. Kizza, M. Uncertainty Assessment in Water Balance Modelling for Lake Victoria. Ph.D. Thesis, Makerere University Kampala, Kampala, Uganda, 2012.

17. Taylor, K.E.; Stouffer, R.J.; Meehl, G.A. An overview of CMIP5 and the experiment design. *Bull. Am. Meteorol. Soc.* **2012**, *93*, 485–498.

18. Taye, M.T.; Ntegeka, V.; Ogiramoi, N.P.; Willems, P. Assessment of climate change impact on hydrological extremes in two source regions of the Nile River Basin. *Hydrol. Earth Syst. Sci.* **2011**, *15*, 209–222.

19. Nyeko-Ogiramoi, P.; Ngirane-Katashaya, G.; Willems, P.; Ntegeka, V. Evaluation and inter-comparison of Global Climate Models' performance over Katonga and Ruizi catchments in Lake Victoria basin. *Phys. Chem. Earth* **2010**, *35*, 618–633.

20. Liu, T.; Willems, P.; Pan, X.L.; Bao, A.M.; Chen, X.; Veroustraete, F.; Dong, Q.H. Climate change impact on water resource extremes in a headwater region of the Tarim basin in China. *Hydrol. Earth Syst. Sci.* **2011**, *15*, 3511–3527.

21. Christensen, J.; Kjellström, E.; Giorgi, F.; Lenderink, G.; Rummukainen, M. Weight assignment in regional climate models. *Clim. Res.* **2010**, *44*, 179–194.

22. Shongwe, M.E.; van Oldenborgh, G.J.; van den Hurk, B.; van Aalst, M. Projected changes in mean and extreme precipitation in Africa under global warming. Part II: East Africa. *J. Clim.* **2011**, *24*, 3718–3733.

23. Ntegeka, V.; Baguis, P.; Roulin, E.; Willems, P. Developing tailored climate change scenarios for hydrological impact assessments. *J. Hydrol.* **2014**, *508*, 307–321.

24. Kizza, M.; Westerberg, I.; Rodhe, A.; Ntale, H.K. Estimating areal rainfall over Lake Victoria and its basin using ground-based and satellite data. *J. Hydrol.* **2012**, *464–465*, 401–411.

25. Shaffrey, L.C.; Stevens, I.; Norton, W.A.; Roberts, M.J.; Vidale, P.L.; Harle, J.D.; Jrrar, A.; Stevens, D.P.; Woodage, M.J.; Demory, M.E.; *et al.* HiGEM: The New U.K. High-Resolution global environment model—Model description and basic evaluation. *J. Clim.* **2009**, *22*, 1861–1896.

26. Hawkins, E.; Sutton, R. The potential to narrow uncertainty in regional climate predictions. *Bull. Am. Meteorol. Soc.* **2009**, *90*, 1095–1107.

27. Ba, M.B.; Nicholson, S.E. Analysis of convective activity and its relationship to the rainfall over the rift valley lakes of East Africa during 1983–90 using the meteosat infrared channel. *J. Appl. Meteorol.* **1998**, *37*, 1250–1264.

28. Shongwe, M.E.; van Oldenborgh, G.J.; Hurk, B. Van Den Projected changes in mean and extreme precipitation in Africa under global warming, Part II: East Africa. *J. Clim.* **2011**, *24*, 3718–3733.

29. ESFG ESFG PCMDI. Available online: http://pcmdi9.llnl.gov/esgf-web-fe/live# (accessed on 31 March 2014).

10

Remote Sensing based Analysis of Recent Variations in Water Resources and Vegetation of a Semi-Arid Region

Shaowei Ning [1,*], **Hiroshi Ishidaira** [1], **Parmeshwar Udmale** [1] **and Yutaka Ichikawa** [2]

[1] International Research Center for River Basin Environment (ICRE), University of Yamanashi, Takeda 4-3-11, Kofu, Yamanashi 400-8511, Japan; E-Mails: ishi@yamanashi.ac.jp (H.I.); pd.udmale@gmail.com (P.U.)

[2] Department of Civil and Earth Resources Engineering, Kyoto University, C1, Kyoto-Daigaku-Katsura, Nishikyo-ku, Kyoto-shi, Kyoto 615-8540, Japan; E-Mail: ichikawa@hywr.kuciv.kyoto-u.ac.jp

* Author to whom correspondence should be addressed; E-Mail: yantaigold@sina.com;

Academic Editor: Yingkui Li

Abstract: This study is designed to demonstrate use of free remote sensing data to analyze response of water resources and grassland vegetation to a climate change induced prolonged drought in a sparsely gauged semi-arid region. Water resource changes over Hulun Lake region derived from monthly Gravity Recovery and Climate Experiment (GRACE) and Tropical Rainfall Measuring Mission (TRMM) products were analyzed. The Empirical Orthogonal Functions (EOF) analysis results from both GRACE and TRMM showed decreasing trends in water storage changes and precipitation over 2002 to 2007 and increasing trends after 2007 to 2012. Water storage and precipitation changes on the spatial and temporal scale showed a very consistent pattern. Further analysis proved that water storage changes were mainly caused by precipitation and temperature changes in this region. It is found that a large proportion of grassland vegetation recovered to its normal state after above average rainfall in the following years (2008–2012) and only a small proportion of grassland vegetation (16.5% of the study area) is degraded and failed to recover. These degraded grassland vegetation areas are categorized as ecologically vulnerable to climate change and protective strategies should be designed to prevent its further degradation.

Keywords: water resource variations; climate change; vegetation; semi-arid region; remote sensing

---

1. Introduction

Freshwater resources are the lifeblood of our planet. It is fundamental to the biochemistry of all living organisms. The Earth's ecosystems are linked and maintained by water; it drives plant growth and provides a permanent habitat for many species, including ourselves. However, freshwater is a resource under considerable pressure. Its stored potential (surface water, ground water, soil moisture, ice, *etc.*) is increasingly facing challenges from climate changes as well as anthropogenic activities. That current and future climate change is expected to significantly impact freshwater systems including rivers, streams and lakes, in terms of flow and direction, timing, availability, temperature, and its inhabitants. So understanding the information about water resource change, its driving force and potential impact in the past and future is very important for water resource management and eco-environmental protection.

In recent years, the response of water resource and vegetation to the changing climate and anthropogenic effects has been discussed extensively at regional or global scales. With the rapid development of remote sensing techniques, the reliability of satellite products relevant to water resource monitoring has greatly improved. For example, changes in terrestrial water storage are measurable through satellite gravity based approximations of equivalent water thickness to a precision of 0.5 cm per year. [1]. Precipitation is monitored by multiple post-processing phases of currently available satellite data (*i.e.*, Tropical Rainfall Measuring Mission (TRMM)) to a resolution of millimeter per day [2]. Water level change in rivers and lakes is derived from altimetry satellites (*i.e.*, Jason-1/2, ENVISAT) to a sub-meter precision [3]. Hence, satellite observations have been increasingly used in such research, exploiting their potential of providing spatially continuous and temporally recurrent estimates over regional to global scales [4].

Zhang *et al.* [5] used monthly precipitation observations over global land areas to analyze precipitation trends in two twentieth century periods (1925–1999 and 1950–1999), and showed that anthropogenic forcing has had a detectable influence on observed changes in average precipitation within latitudinal bands, and that these changes cannot be explained by internal climate variability or natural forcing. Syed *et al.* [6] characterized terrestrial water storage variations using Gravity Recovery and Climate Experiment (GRACE) and Global Land Data Assimilation System (GLDAS) at global scale, the results illustrated spatial-temporal variability of water storage change over land, with implications for a better understanding of how terrestrial water storage responds to climate change and variability. Apart from global scale studies, Fensholt and Du [7,8] assessed the regional/continental precipitation trends and showed their influence on stream flow, water level, soil moisture and vegetation changes. Moiwo *et al.* [9] analyzed water storage dynamics in the North China Region (an important grain-production base) using GRACE, GLDAS products in conjunction with *in situ* hydro-climate data, the results showed a sharp water storage depletion from April 2002 through December 2009 in that area and water loss which was more a human than a natural cause had already negatively influenced millions of people in the region and beyond in terms of water supply crop production, eco-environmental system and social stability.

Besides that, much research also indicates that a remote sensing approach is a cost-efficient and accurate method to monitor inland water surface and water level (case of lakes and reservoirs) dynamics which are also affected directly by climate change and human activity. Dorothea *et al.* [10] used a Moderate Resolution Imaging Spectro-radiometer (MODIS) surface reflectance dataset and a Modified Normalized Difference Water Index (*MNDWI*) to map the variability of Lake Manyara's water surface area over 2000–2011. Their results implied that recent fluctuations of Lake Manyara's surface water area are a direct consequence of global and regional climate fluctuations. Duan *et al.* [11] proposed and evaluated a method that combined operational satellite altimetry databases with satellite imagery data to estimate water volume variations in Lake Tana. Results showed that satellite altimetry products were in good agreement with *in situ* water levels for Lake Tana ($R^2 = 0.97$). Estimated water volume variations derived from satellite altimetry products and LANDSAT TM/ETM+ agreed well with *in situ* water volume for Lake Tana, with R^2 higher than 0.95 and Root Mean Square Error (RMSE) 9.41% of corresponding mean value of *in situ* measurements.

With respect to vegetation, it plays a notably important role in soil conservation, atmosphere adjustment and maintenance of climatic and whole ecosystem stability because of its natural tie connecting atmosphere, water, and soil. Surface vegetation conditions are known for their sensitivity to natural changes and anthropogenic effects, thus serving as important proxies for regional eco-environmental and global climate fluctuations. Satellite based vegetation indexes such as normalized difference vegetation index (*NDVI*) as an efficient tool are widely used to examine the dynamic of vegetation health, density and degradation due to climate changes and anthropogenic effects [12,13].

As mentioned above, satellite remote sensing has shown promising results in the estimation of water resources and vegetation. However, in this study, we focus on the analysis of a combination of available satellite data including GRACE terrestrial water storage (TWS), TRMM, MODIS/LANDSAT, satellite altimetry data (Topex/Poseidon, Jason-1/2) coupled with *in situ* climate data to assess the water resource variation within a sparsely gauged area—the Hulun lake region and its impact on the eco-environment to provide useful information for future water resource management and eco-environmental protection. More specifically, this study aims (1) to provide a framework for a remote sensing based integrated assessment of water resource trends; (2) to detect trends in consistently established time series (from 2002 to 2012) of terrestrial water storage change and precipitation in a spatial distributed manner and (3) to infer the probable causes of water resource variations and its impacts on vegetation in order to contribute towards sustainable eco-environmental management.

2. Study Area

For this study, a representative case of the Daurian Steppe Eco-region (a most intact example of Eurasian Steppe) is selected (Figure 1). It is straddled over borders of three countries, namely, China, Mongolia and Russia (111° E–119° E, 47° N–50° N). The total study area is about 290,400 km². It covers a part of an ecologically important region—The Daurian International Protected Areas (DIPA), namely, The Hulun Lake Nature Reserve grassland. The Hulun Lake Nature Reserve is a reserved grassland and least influenced by human activities [14,15]. This draws attention to identify the consequences of water resource changes (consecutive years of precipitation deficit and decline in TWS) on representative natural grassland-vegetation with minimum anthropogenic disturbances.

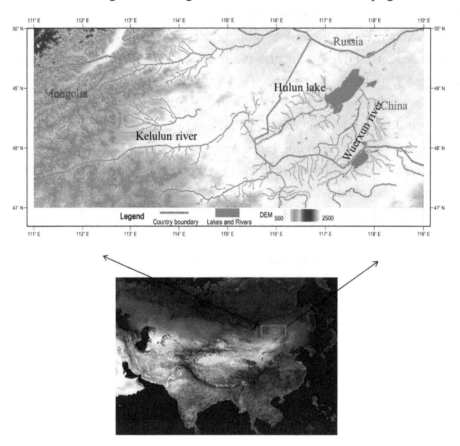

Figure 1. The geographic location and Shuttle Radar Topography Mission (SRTM)-based elevation map of the study area (red color region in the down figure shows the Eurasian Steppe zone).

The area has a mid-temperate semi-arid continental climate with the dominant mid-temperate zone characterized by drastic changes in winter and summer seasons. The average annual rainfall is about 293 mm, mainly concentrated in unfrozen season (from May to October). The average annual temperature ranges from −13 °C in winter to 12.3 °C in summer, the average annual evaporation is around 249 mm, and the average annual relative humidity is 49%. The semi-arid climate with the strong winds is increasing the vulnerability of this area to desertification. Figure 2 shows the vegetation cover types of the study area. About 89.9% and 7.9% of the study area is occupied by annual grass-vegetation and forest (deciduous needle leaf, deciduous broadleaf, evergreen needle leaf and annual broadleaf vegetation), respectively. The surface water bodies cover about 2.2% of study area, with a major water bodies—Hulun Lake having surface area 2307 km^2 and Beier Lake with surface area 609 km^2. As shown in Figure 1, there are many inland rivers in the study area, but only two rivers, Kelulun and Wuerxun river, with annual discharge about 7×10^8 and 5.5×10^8 m^3, flow into Hulun lake which is the main drainage outlet in this area. Recently, however, the annual average discharge of these two rivers is less than 2×10^8 m^3. The fluctuations in water levels of Hulun lake can be used as an indicator of wet and dry conditions in the study area.

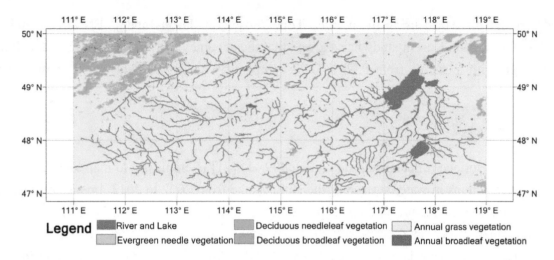

Figure 2. Vegetation map of study area.

3. Materials

3.1. Precipitation Data

The monthly precipitation for the period of 2002–2012 is obtained from Tropical Rainfall Measuring Mission (TRMM) [16]. TRMM products have been used in a number of studies of Inner Mongolia and surrounding precipitation, where they have been found to be adequate when compared with ground observations [17,18]. The product used in this study is referred as the TRMM and other precipitation dataset (denoted as 3B43). It is derived not only from TRMM sensors but also a number of other satellites and ground based rain gauged data. Monthly observed precipitation data (2002–2012) for five stations near the Hulun Lake are employed in this analysis to evaluate the applicability of satellite derived precipitation in study area.

3.2. Surface Air Temperature Data

This study uses a surface air temperature dataset namely Global Historical Climatology Network version 2 and the Climate Anomaly Monitoring System (GHCN + CAMS) which is a station observation-based global land monthly mean surface air temperature dataset at 0.5 × 0.5 degree resolution for the period of 1948 to the present. When compared with several existing observation-based land surface air temperature data sets, the preliminary results show that the quality of this new GHCN + CAMS land surface air temperature analysis is reasonably good and the new dataset can capture most common temporal-spatial features in the observed climatology and anomaly fields over both regional and global domains [19].

3.3. Lake Water Level Data

Monthly water level data for the Hulun Lake for the period of 2002–2012 is obtained from Hydroweb dataset [20]. The dataset is developed by Laboratoire d'Etudes en Oceanographie et Geode'sie Spatiale, Equipe Geodesie, Oceanographie, et Hydrologie Spatiales (LEGOS/GOHS) in Toulouse, France. It provides time series of water levels of large rivers, about 150 lakes and reservoirs, and wetlands around the world using the merged Topex/Poseidon, Jason-1and 2, ENVISAT and Geosat Fellow-On (GFO) data. Recent study has showed that the accuracy of water level data from Hydroweb was very high with R^2 range from 0.96 to 0.99 compared with *in situ* data in US, Netherlands and Ethiopia [11].

3.4. Satellite Imagery Data

MODIS Terra surface reflectance product (Mod09A1) [21] is employed to map and monitor spatial and temporal variations in water surface of the Hulun Lake from 2002 to 2012. Images on which snow covers the lake surface and surrounding region in winter time are not selected, because it is difficult to retrieve lake surface area in those scenes. Besides that, we also use several scenes of LANDSAT TM/ETM+ data [22] with spatial resolution of 30m to validate lake surface area derived from Mod09 A1.

3.5. GRACE TWS Data

TWS was derived from the latest version monthly GRACE gravity solutions (RL05) generated by the Center for Space Research at the University of Texas at Austin [23], from August 2002 through December 2012. Each solution consists of sets of spherical harmonic (Stokes) coefficients, C_{lm} and S_{lm}, to degree 1 and order m, both size less than or equal to 60. We calculated these coefficients by combining GRACE data with ocean model output as Swenson *et al.* [24] did. TWS calculation and the post processing method used here were similar with Duan *et al.* [25] with two Fan filter [26] radiuses 500 and 800 km, respectively. Finally, these coefficients were transformed into 1×1 degree gridded data that reflect vertically integrated water mass change represented by equivalent water thickness.

3.6. NDVI Data

NDVI dataset acquired from the Advanced Very High Resolution Radiometer (AVHRR) sensor aboard NOAA satellites processed by the Global Inventory Monitoring and Modeling Studies (GIMMS) at the National Aeronautics and Space Administration (NASA) [27]. The database ranges from July 1981 to December 2013 at a spatial resolution 8 km^2. The data are composited over approximately 15 day periods with the maximum value compositing technique, which minimizes the influences of atmospheric aerosols and clouds. This study analyzes the *NDVI* trend for the period from 2002 to 2012.

4. Methods

4.1. Water Resource Spatial-Temporal Series Analysis

Empirical Orthogonal Functions (EOF) analysis is a widely and easily used statistical method for analyzing large multidimensional datasets. When applied to a space-time dataset, EOF analysis can be used to decompose the observed variability into a set of spatial change patterns (EOFs), which are statistically independent and spatially orthogonal to the others, and a set of times series called time coefficients (PCs) that describes the time evolution of the particular EOFs. Together, the EOFs and PCs can be combined to reconstruct the variability in the original dataset. Basically, the goal of EOF analysis is to transform an original set of variables into a substantially smaller set of uncorrelated variables, which can reflect most of the information of the original dataset. It also has the ability to isolate various processes mixed in observation data [28]. The EOF has recently become a popular tool in various science areas such as meteorology, geology, and geography [29]. In this study, EOF analysis is applied to study both the spatial and temporal changes of precipitation and TWS.

4.2. Lake Water Surface Area Estimation

Several land cover classification methods can be used for delineating water bodies from multi-temporal satellite imagery to date from conventional unsupervised methods to more advanced artificial neural networks and support vector machine classifier [12,30]. The Modified Normalized Difference Water Index (*MNDWI*) method proposed by Xu [31] has been widely applied and proved efficient to retrieve water surface. The *MNDWI* is a band ratio index between Green (correspond to band 4 of MOD09A1 imagery) and Shortwave Infrared (*SWIR*, correspond to band 6 of MOD09A1 imagery) spectral bands that enhances water features. MNDWI is defined as:

$$MNDWI = \frac{Green - SWIR}{Green + SWIR} \tag{1}$$

Following other studies [32,33], we set the threshold for *MNDWI* to zero. *MNDWI* values > 0 represent water bodies and < 0 non-water cover types. Water features have positive *MNDWI* values because of their higher reflectance in the Green band than in the SWIR band while non-water features (soil and vegetation) have negative *MNDWI* values due to their low reflectance in the Green band than the SWIR band. However, some parts of Hulun Lake with the average depth 5.7 m are very shallow, usually less than 1 m and many aquatic plants grow out of water surface. This makes the *MNDWI* values negative in some grid located inside the lake. Here, we combine the *NDVI* value to eliminate that effect. We decide if *NDVI* < 0 or *MNDWI* > 0 and only one rule satisfies, then it is classified as water body. To validate the results, we estimate the water body from several scenes of LANDSAT imagery by traditional manual digitization, which is time consuming but has high accuracy.

4.3. NDVI Variation Trend Analysis Method

The Theil-Sen Median trend analysis, Mann-Kendall [34] are used to study the vegetation covered regions of our study area, namely, the temporal variation characteristics of the *NDVI* of the pixel covered region with *NDVI* values greater or equal to 0.1. The Theil-Sen trend analysis method can be effectively combined with the Mann-Kendall test. These are important methods for detecting the trend of long time series data, and this combination has been gradually used to analyse the long time series of vegetation reflecting the variation in trends of each pixel in a time series.

The Theil-Sen Median trend analysis is a robust trend statistical method, and it calculates the median slopes between all $n \cdot (n - 1)/2$ pair-wise combinations of the time series data. It is based on non-parametric statistics and is particularly effective for the estimation of trends in small series. The slope of Theil-Sen Median can represent the increase or decrease in the *NDVI* over the 11 years between 2002 and 2012 on a pixel scale. It is calculated by:

$$TS_{NDVI} = median\left(\frac{NDVI_m - NDVI_n}{m - n}\right), 2002 \leq n < m \leq 2012 \tag{2}$$

where, TS_{NDVI} refers to the Theil-Sen median, and $NDVI_m$, $NDVI_n$ represent the NDVI values for years of m and n, in case of $TS_{NDVI} > 0$, the NDVI shows a rising trend, otherwise, the NDVI presents a decreasing trend.

The Mann-Kendall test measures the significance of a trend. It is a non-parametric statistical test, and it has the advantage that samples do not need to follow certain distributions and is free from the interference of outliers. It has been broadly used to analyse the trends and variations at sites with hydrological and meteorological time series. Recently, this method has been applied to detection of vegetation variations over long time periods. The calculation algorithm is as follows:

It is assumed that $NDVI_m$, m stands for time series from 2002 to 2012. The statistics of Z is defined as:

$$Z = \begin{cases} \dfrac{S-1}{\sqrt{s(S)}} & ,s > 0 \\ 0 & ,s = 0 \\ \dfrac{S+1}{\sqrt{s(S)}} & ,s < 0 \end{cases} \tag{3}$$

where, $S = \sum_{n=1}^{t-1} \sum_{m=n+1}^{t} sgn(NDVI_n - NDVI_m)$,

$$sgn(NDVI_n - NDVI_m) = \begin{cases} 1 & ,NDVI_n - NDVI_m > 0 \\ 0 & ,NDVI_n - NDVI_m = 0 \\ -1 & ,NDVI_n - NDVI_m < 0 \end{cases},$$

$$s(s) = \frac{t(t-1)(2t+5)}{18} \tag{4}$$

where, $NDVI_m$ and $NDVI_n$ stands for the NDVI values of the pixels m and n; t is the length of the time series; sgn is a sign function; and the Z statistic is located in the range of $(-\infty, +\infty)$. A given significance level, $|Z| > \mu_{1-\alpha/2}$, signifies that the times series shows significant variations on the level of α. Generally, the value of α is 0.05. In this study, we choose $\alpha = 0.05$, means that we measure the significance of the NDVI trend over period from 2002 to 2012 on pixel scale at a confidence level of 0.05.

5. Results and Discussion

5.1. Precipitation and Temperature Variation Analysis

The numbers of rain-gauge stations in the study area are limited. Hence, we used TRMM monthly data to analyze precipitation trends in this study. To confirm the feasibility of TRMM data, observed precipitation from five rain gauge stations (Table 1) located in the vicinity of the study area are used. The correlation between two data sets for respective grids is observed to be in the range of 0.74–0.94 over the period of 2002–2012 as shown in Table 1. This confirms the applicability of TRMM data for precipitation trend analysis in this study. Studies by Yatagai et al. [17] and Chen et al. [18] also validated the applicability of TRMM data in this region.

After applying EOF to TRMM data, we found three dominant EOFs and PCs in study area (as shown in Figure 3). EOF1 and PC1 represent about 65% of total variance of precipitation, which shows superposition of annual and seasonal variability. The EOF1 is found positive throughout the study area with high values in central part highlighting uniform changing pattern over study area.

A significant decreasing trend is observed over a period of 2002–2007, however an increasing trend from 2008 to 2012 can be seen from PC1. Low negative PC1 values corresponding to summer season from 2003 to 2007 indicates five consecutive years of below average rainfall, which induced a very serious drought. Using a linear regression, we found an average precipitation decline of 23.1 mm/year and increase of 18.2 mm/year for the periods of 2002–2007 and 2008–2012, respectively. We do not interpret the second and third mode of EOF on precipitation variation (*i.e.*, EOF2, PC2 and EOF3, PC3 here, respectively), since the temporal pattern change is not obvious, and it accounts for only 13% and 6% of variance in precipitation, respectively.

Table 1. Correlation between gauged precipitation with tropical rainfall measuring mission (TRMM) precipitation in respective grids.

No.	Station Name	Latitude	Longitude	R^2
1	Xinyouqi	48.67° N	116.82° E	0.75
2	Xinzuoqi	48.21° N	118.27° E	0.74
3	Manzhouli	49.57° N	117.43° E	0.82
4	Hailaer	49.22° N	119.75° E	0.92
5	Aershan	47.17° N	119.93° E	0.94

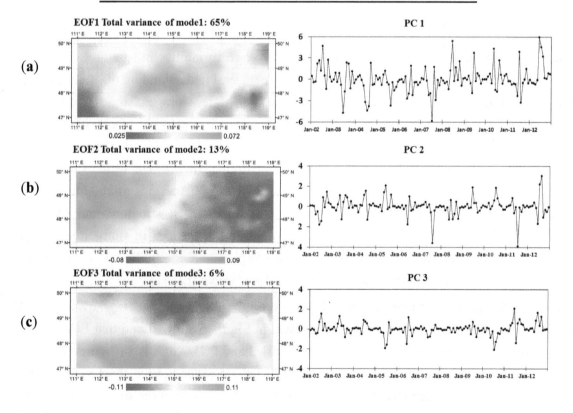

Figure 3. EOF decomposition of precipitation changes derived from TRMM satellite data over study area. EOF patterns are shown in left side and corresponding unit-less temporal patterns (PCs) in right side. (**a**): The first change mode of precipitation changes; (**b**): The second change mode of precipitation changes; (**c**): The third change mode of precipitation changes.

We also analyzed the warm (May–October) and cold (November–April) season average temperature over study area as shown in Figure 4. Average temperature of warm and cold season shows opposite change pattern against precipitation. Rising temperature may have caused more evapotranspiration, then further exacerbated water storage depletion and drought.

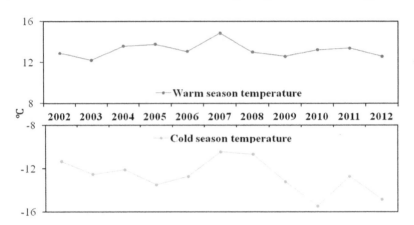

Figure 4. Temperature trend over study area from 2002 to 2012.

5.2. Water Storage Change

Figure 5 shows the overall EOF results of water storage variability in spatial and temporal scale over the study area. EOF analysis result provides us the general understanding of water storage change condition. EOF1 and PC1 represent 88% of total variance in water storage changes and EOF2 and PC2 represent only 10%. For EOF1, all values are positive, which means all places have same change pattern with high significance in the central and north part of study area, while the corresponding PC1 shows the dominant trend. By a linear regression, we found an average water storage decline of 14.2 mm/year and an increase of 3 mm/year in the study area for the periods of 2002 to 2007 and 2008 to 2012, respectively. EOF2 delineates a spatial east-west dipole structure and PC2 shows an increasing trend in the southwest corner and decreasing trend in the east of study area over period of 2008 to 2012. Overall, the water storage first reduced sharply (2002 to 2007) and then restored slightly (2008–2012), especially in the central and north part of study area (EOF1).

EOF2 and PC2 (Figure 3) representing 13% of total rainfall change pattern shows the similar east west dipole structure of the TWS pattern 2 (as shown in Figure 5). It is found that EOF1/2 and PC1/2 of precipitation (Figure 3) is very consistent with EOF1/2 and PC1/2 of TWS (Figure 5). This explains that the precipitation is one of the major driving factors behind water storage changes. Similar trends in precipitation and water storage changes observed over whole study area (as shown in Figure 6), which indicates a very sharp decreasing trend over the period of 2002–2007. However, for the period 2007–2012, in spite of the increasing precipitation trend, TWS did not show a significant increasing trend as that of precipitation but increased slightly. There may be two reasons for that: first, as we mentioned in the study area section, water income from the two rivers flow to this region has been lower than usual recently; second, actual evapotranspiration has increased because of above-normal vegetation development (as we will explain in Section 5.4).

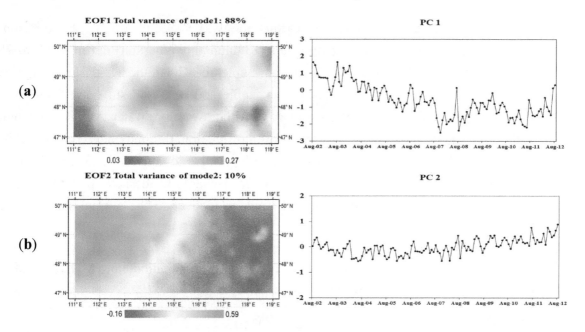

Figure 5. EOF decomposition of TWS changes over the study area. EOF patterns are shown on the left side and corresponding unit-less temporal patterns (PCs) are shown on the right side. (**a**): The first change mode of TWS changes; (**b**): The second change mode of TWS changes.

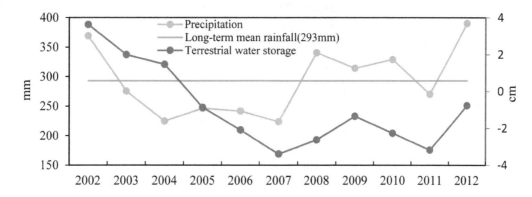

Figure 6. Annual precipitation and water storage changes series from 2002 to 2012.

5.3. Lake Response to Water Storage Change

As mentioned in the study area introduction, Lake Hulun drains in most parts of the study area, so its water volume change is a good indicator of water resource variation for this region. Table 2 shows the comparison between MODIS derived and LANDSAT digitalized results of lake surface area. It shows a very close relation with R^2 value as high as 0.95 and very low absolute and relative errors. It demonstrates the validity of MODIS derived lake surface area with high accuracy in unfrozen season (May–October). Figure 7 gives the lake level and lake water area time series with their correlation. Both shows rapid declination from 2002 to 2009 with about three meters of water level drop and 400 km² of shrinkage in lake surface area, respectively, and remained stable after 2009. This temporal change pattern is not consistent with water storage and precipitation change. Although precipitation increased after 2007, lake volume still decreased till 2009 and did not show obvious rise after that. For this phenomenon, there may be several possible reasons. Firstly, because of drought, Lake Hulun got less inflow from the upper reaches of two main rivers (Kelulun and Wuerxun) during 2002 to 2007. According to local news, these two rivers dried up from September 2007 and did not discharge water into the lake for almost for one year [35]. Secondly, people and livestock suffering from drought, which has driven water scarcity around the lake may have withdrawn more water from it than a normal year. Finally, the most important point, a large proportion of precipitation may have been contributed to recover soil moisture deficit and depleted groundwater levels due to consecutive years of droughts. This might have delayed the river inflow and groundwater discharge to the lake. In summary, when drought attacks this region, it needs more time and water to recover to the normal state even after enough rainfall.

Table 2. Comparison of Moderate Resolution Imaging Spectro-radiometer (MODIS) and LANDSAT derived Hulun lake area.

LANDSAT		MODIS		Absolute Error (km²)	Relative Error (%)
Date	Lake Surface Area (km²)	Date	Lake Surface Area (km²)		
1 July 2000	2306.6	4 July 2000	2290.1	−16.4	−0.71%
6 September 2001	2236.4	7 September 2001	2186.3	−50.2	−2.24%
8 August 2002	2154.4	6 August 2002	2221.2	66.7	3.10%
27 August 2003	2114.2	30 August 2003	2106.9	−7.3	−0.35%
13 August 2004	2002.5	13 August 2004	2058.2	55.7	2.78%
16 August 2005	1977.6	14 August 2005	1948.8	−28.7	−1.45%
26 July 2006	1938.6	29 July 2006	1942.9	4.3	0.22%
29 July 2007	1907.2	29 July 2007	1902.3	−4.9	−0.26%
25 August 2008	1837	13 August 2008	1862.1	25.1	1.36%
5 October 2009	1791.5	1 October 2009	1772.8	−18.7	−1.04%

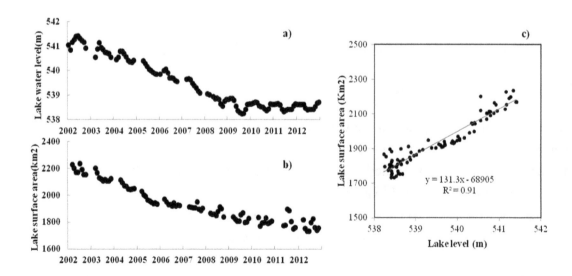

Figure 7. (**a**) Lake water level; (**b**) lake water surface area; and (**c**) correlation between lake water level and surface area.

5.4. Vegetation Response to Water Resource Change

Average *NDVI* distribution over the period of 2002–2012 is shown in Figure 8a. It can be seen that the areas with high *NDVI* (0.4–0.5) are located in northwestern part. *NDVI* values in central and south part of study area, and in the vicinity of lake grassland region are relatively low (0.2–0.3). The variations in trends of *NDVI* can be effectively captured by the Theil-Sen median trend analysis and the Mann-Kendall test to reflect the spatial distribution of vegetation responses to water resource changes. Because regions with a TS_{NDVI} of 0 strictly do not exist, we made the following classifications according to the real conditions of the TS_{NDVI}. Regions with a TS_{NDVI} from −0.0005 to 0.0005 are categorized as stable regions, regions with TS_{NDVI} larger than or equal to 0.0005 are categorized improved regions and regions with TS_{NDVI} less than −0.0005 are categorized as degraded areas. Moreover, significance test results of the Mann-Kendall test, at the confidence level of 0.05, are determined as significance variations ($Z > 1.96$ or $Z < -1.96$) or insignificant variations ($-1.96 \leq Z \leq 1.96$). Through combining the classification results of the Theil-Sen median trend analysis and the MK test, it is comparable with the data of trend variations of the *NDVI*. The results are summarized into five classes as shown in Table 3. It shows the regions with vegetation condition improvement, regions with stable vegetation condition, and regions with vegetation degradation, which account for 70%, 13.5% and 16.5%, respectively.

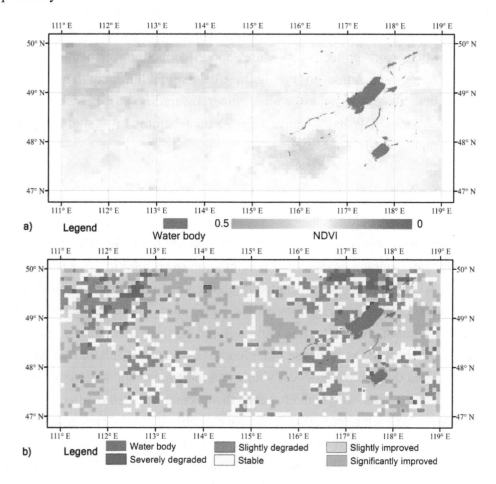

Figure 8. Spatial distribution of (**a**) average *NDVI* and (**b**) trends of inter-annual *NDVI* change over 2002 to 2012.

Table 3. Changing trend of normalized difference vegetation index (NDVI) in the study area.

TS_{NDVI}	Z	NDVI Trend	Area (%)
<-0.0005	<-1.96	Severely degraded	2.69%
<-0.0005	$-1.96-1.96$	Slightly degraded	13.80%
$-0.0005-0.0005$	$-1.96-1.96$	Stable	13.51%
≥ 0.005	$-1.96-1.96$	Slightly improved	57.35%
≥ 0.005	≥ 1.96	Improved	12.64%

As shown in Figure 8b, the region with improved vegetation condition is far larger than the regions with degrading trend, and mainly scattered in the central part of the study area. This indicates resilience of grassland vegetation to droughts. The decreasing precipitation from 2002 to 2007 had not much impact on vegetation in the central part. It can be seen that the vegetation condition recovered quickly after the withdrawal of drought and the increase in precipitation in the following years. Hence, average *NDVI* over 2002–2012 showed an increasing trend. This shows the close relationship between vegetation conditions and precipitation as found by other researchers [36]. It is obvious to detect such vegetation conditions with a decreasing trend distributed in the northwest mountain areas and Lake Hulun surrounding areas (especially, in the northern and western part, Figure 2 shows that these are the evergreen and deciduous forest areas). Decrease in water resources had serious negative and long-term impacts on vegetation conditions. Even the precipitation increased after 2007, vegetation conditions did not recover to its normal state. We can categorize those areas as ecologically vulnerable regions, where more protective measures and effective management are needed. Possible causes about vegetation degradation in the north of Lake Hulun may have also affected by anthropogenic activities. Since it became the more important trading port between China and Russia after 2000, human activities have been more frequent than before. In addition, rapid urbanization affected vegetation conditions in this area.

6. Conclusions

In the present study, water resource changes over Hulun Lake region derived from monthly GRACE and TRMM products were analyzed. The EOF results from both GRACE and TRMM showed decreasing trends in water storage changes and precipitation over 2002 to 2007 and increasing trends after 2007 to 2012. Water storage and precipitation changes in spatial and temporal scale showed a very consistent pattern. Further analysis proved that water storage changes were mainly caused by precipitation and temperature changes in this region. Based on the general understanding about water resource variations, we checked the response of Hulun Lake. Results indicated that lake level and lake surface area both declined during 2002 to 2009, with about three meters of water level drop and 400 km^2 shrinkage in lake surface area, respectively, and then remained stable after 2009 even though precipitation had recovered back to pre-2002 level. We can infer that water resource conditions needed more time and precipitation to recover from a long term drought in this typical semi-arid region. Furthermore, the vegetation response to water resource variations reflected that vegetation resilience to drought in most regions was high, forests were less resilient to drought than grasslands. Drought did not bring serious negative implications on vegetation growing conditions. Only 16.5% of the study area which is located in the northern and western sections of Hulun Lake and northwest mountain areas showed vegetation degradation. These areas that are categorized as ecological vulnerable regions need more protection and effective management in the future. Finally, this study demonstrated the feasibility of estimating water resource variation on the spatial–temporal scale and its impact on eco-environment using freely available remote sensing data in a sparsely gauged semi-arid area, which can also be adapted to other regions. Such spatiotemporally distributed analysis at the regional and basin level is particularly important considering that most of the water management and eco-environmental protection also take place at these scales.

Acknowledgments

The authors would like to express their sincere gratitude to the Ministry of Education, Culture, Sports, Science and Technology (Monbukagakusho: MEXT) and University of Yamanashi, Japan for providing financial assistance to undertake this study.

Author Contributions

Then manuscript was primarily written by Shaowei Ning, with Parmeshwar Udmale contributing to its preparation and English check. Hiroshi Ishidaira and Yutaka Ichikawa supervised the research and critically reviewed the draft.

References

1. Long, D.; Scanlon, B.R.; Longuevergne, L.; Sun, A.-Y.; Fernando, D.N.; Save, H. GRACE satellites monitor large depletion in water storage in response to the 2011 drought in Texas. *Geophys. Res. Lett.* **2013**, *40*, 3395–3401.

2. Huffman, G.J.; Bolvin, D.T.; Nelkin, E.J.; Wolff, D.B.; Adler, R.F.; Gu, G.; Yang, H.; Kenneth, P.B.; Erich, F.; Stocker, E.F. The TRMM multi-satellite precipitation analysis (TMPA): Quasi-global, multiyear, combined-sensor precipitation estimates at fine scales. *J. Hydrometeorol.* **2007**, *8*, 38–55.

3. Wang, X.; Gong, P.; Zhao, Y.; Xu, Y.; Cheng, X.; Niu, Z.; Luo, Z.; Huang, H.; Sun, F.; Li, X. Water-level changes in China's large lakes determined from ICESat/GLAS data. *Remote Sens. Environ.* **2013**, *32*, 131–144.

4. Gokmen, M.; Vekerdy, Z.; Verhoef, W.; Batelaan, O. Satellite based analysis of recent trends in the ecohydrology of a semi-arid region. *Hydrol. Earth Syst. Sci. Discuss.* **2013**, *10*, 6193–6235.

5. Zhang, X.; Zwiers, F.W.; Hegerl, G.C.; Lambert, F.H.; Gillett, N.P.; Solomon, S.; Nozawa, T. Detection of human influence on twentieth-century precipitation trends. *Nature* **2007**, *448*, 461–465.

6. Syed, T.H.; Famiglietti, J.S.; Rodell, M.; Chen, J.; Wilson, C.R. Analysis of terrestrial water storage changes from GRACE and GLDAS. *Water Resour. Res.* **2008**, *44*, doi:10.1029/2006WR005779.

7. Fensholt, R.; Rasmussen, K. Analysis of trends in the Sahelian "rain-use efficiency" using GIMMS NDVI, RFE and GPCP rainfall data. *Remote Sens. Environ.* **2011**, *115*, 438–451.

8. Du, J.; He, F.; Zhang, Z.; Shi, P. Precipitation change and human impacts on hydrologic variables in Zhengshui River Basin, China. *Stoch. Environ. Res. Risk Assess.* **2011**, *25*, 1013–1025.

9. Moiwo, J.P.; Tao, F.; Lu, W. Analysis of satellite-based and in situ hydro-climatic data depicts water storage depletion in North China Region. *Hydrol. Process.* **2012**, *27*, 1110–1020.

10. Dorothea, D.; Richard, G. Remote sensing analysis of lake dynamics in semi-arid regions: Implication for water resource management. *Water* **2013**, *5*, 698–727.

11. Duan, Z.; Bastiaanssen, W.G.M. Estimating water volume variations in lakes and reservoirs from four operational satellite altimetry databases and satellite imagery data. *Remote Sens. Environ.* **2013**, *134*, 403–416.

12. Eckert, S.; Hüsler, F.; Liniger, H.; Hodel, E. Trend analysis of MODIS NDVI time series for detecting land degradation and regeneration in Mongolia. *J. Arid Environ.* **2015**, *113*, 16–28.

13. Jiang, W.; Yuan, L.; Wang, W.; Cao, R.; Zhang, Y.; Shen, W. Spatio-temporal analysis of vegetation variation in the Yellow River Basin. *Ecol. Indic.* **2015**, *51*, 117–126.

14. Shinoda, M.; Nachinshonhor, G.U.; Nemoto, M. Impact of drought on vegetation dynamics of the Mongolian steppe: A field experiment. *J. Arid Environ.* **2010**, *74*, 63–69.

15. Da Silva, E.C.; de Albuquerque, M.B.; de Azevedo Neto, A.D.; da Silva Junior, C.D. Drought and its consequences to plants—From individual to ecosystem. In *Responses of Organisms to Water Stress*; InTech: Rijeka, Croatia, 2013.

16. Kummerow, C.; Barnes, W.; Kozu, T.; Shiue, J.; Simpson, J. The tropical rainfall measuring mission (TRMM) sensor package. *J. Atmos. Ocean. Technol.* **1998**, *15*, 809–817.

17. Yatagai, A.; Xie, P.; Kitoh, A. Utilization of a new gauge-based daily precipitation dataset over monsoon Asia for validation of the daily precipitation climatology simulated by the MRI/JMA 20-km-mesh AGCM. *SOLA* **2005**, *1*, 193–196.

18. Chen, Y.; Velicogna, I.; Famiglietti, J.S.; Randerson, J.T. Satellite observations of terrestrial water storage provide early warning information about drought and fire season severity in the Amazon. *J. Geophys. Res. Biogeosci.* **2013**, *118*, 495–504.

19. Fan, Y.; van den Dool, H. A global monthly land surface air temperature analysis for 1948–present. *J. Geophys. Res.* **2008**, *113*, doi:10.1029/2007JD008470.

20. LEGOS/GOHS. Available: http://www.legos.obs-mip.fr (accessed on 3 January 2013).

21. U.S. Government Computer. Available online: http://e4ftl01.cr.usgs.gov/MOLT/MOD09A1 (accessed on 14 March 2015).

22. Landsat Data Access. Available online: http://landsat.usgs.gov/Landsat_Search_and_Download.php (accessed on 20 March 2015).

23. Center for Space Research. Available online: ftp://podaac.jpl.nasa.gov/allData/grace/L2/CSR/RL05/ (accessed on 12 February 2013).

24. Swenson, S.; Chambers, D.; Wahr, J. Estimating geocenter variations from a combination of GRACE and ocean model output. *J. Geophys. Res.* **2008**, *113*, doi:10.1029/2007JB005338.

25. Duan, X.J.; Guo, J.Y.; Shum, C.K.; Wal, W. On the postprocessing removal of correlated errors in GRACE temporal gravity field solutions. *J. Geod.* **2009**, *83*, 1095–1106.

26. Zhang, Z.-Z.; Chao, B.F.; Lu, Y.; Hsu, H.-T. An effective filtering for GRACE time-variable gravity: Fan filter. *Geophys. Res. Lett.* **2009**, *36*, doi:10.1029/2009GL039459.

27. GIMMS, "ECOCAST", NASA. Available online: http://ecocast.arc.nasa.gov/data/pub/gimms/ (accessed on 19 April 2015).

28. Longuevergne, L.; Florsch, N.; Elsass, P. Extracting coherent regional information from local measurements with Karhunen-Loève transform: Case study of an alluvial aquifer (Rhine valley, France and Germany). *Water Resour. Res.* **2007**, *43*, doi:10.1029/2006WR005000.

29. Perry, M.A.; Niemann, J.D. Analysis and estimation of soil moisture at the catchment scale using EOFs. *J. Hydrol.* **2007**, *334*, 388–404.

30. Song, X.; Duan, Z.; Jiang, X. Comparison of artificial neural networks and support vector machine classifiers for land cover classification in Northern China using a SPOT-5 HRG image. *Int. J. Remote Sens.* **2012**, *33*, 3301–3320.

31. Xu, H. Modification of normalised difference water index (NDWI) to enhance open water features in remotely sensed imagery. *Int. J. Remote Sens.* **2006**, *27*, 3025–3033.

32. Deus, D.; Gloaguen, R. Remote Sensing Analysis of Lake Dynamics in Semi-Arid Regions: Implication for Water Resource Management, Lake Manyara, East African Rift, Northern Tanzania. *Water* **2013**, *5*, 698–727.

33. Cai, B.F.; Yu, R. Advance and evaluation in the long time series vegetation trends research based on remote sensing. *J. Remote Sens.* **2009**, *13*, 1170–1186.

34. Hoaglin, D.C.; Mosteller, F.; Tukey, J.W. *Understanding Robust and Exploratory Data Analysis*; Wiley: New York, NY, USA, 1983.

35. National Center for Agricultural Scientific Data Sharing. Available online: http://grassland. agridata.cn/client.c?method=fgbz&menuid=106&id=59 (accessed on 26 May 2015).

36. Ding, M.; Zhang, Y.; Liu, L.; Zhang, W.; Wang, Z.; Bai, W. The relationship between *NDVI* and precipitation on the Tibetan Plateau. *J. Geogr. Sci.* **2007**, *17*, 259–268.

11

Suitability of a Coupled Hydrodynamic Water Quality Model to Predict Changes in Water Quality from Altered Meteorological Boundary Conditions

Leon van der Linden *, Robert I. Daly and Mike D. Burch

South Australian Water Corporation, Adelaide, SA 5000, Australia;
E-Mails: rob.daly@sawater.com.au (R.I.D.); mike.burch@sawater.com.au (M.D.B.)

*   Author to whom correspondence should be addressed; E-Mail: leon.vanderlinden@sawater.com.au;

Academic Editor: Julia Piantadosi

Abstract: Downscaled climate scenarios can be used to inform management decisions on investment in infrastructure or alternative water sources within water supply systems. Appropriate models of the system components, such as catchments, rivers, lakes and reservoirs, are required. The climatic sensitivity of the coupled hydrodynamic water quality model ELCOM-CAEDYM was investigated, by incrementally altering boundary conditions, to determine its suitability for evaluating climate change impacts. A series of simulations were run with altered boundary condition inputs for the reservoir. Air and inflowing water temperature (TEMP), wind speed (WIND) and reservoir inflow and outflow volumes (FLOW) were altered to investigate the sensitivity of these key drivers over relevant domains. The simulated water quality variables responded in broadly plausible ways to the altered boundary conditions; sensitivity of the simulated cyanobacteria population to increases in temperature was similar to published values. However the negative response of total chlorophyll-a suggested by the model was not supported by an empirical analysis of climatic sensitivity. This study demonstrated that ELCOM-CAEDYM is sensitive to climate drivers and may be suitable for use in climate impact studies. It is recommended that the influence of structural and parameter derived uncertainty on the results be evaluated. Important factors in determining phytoplankton growth were identified and the importance of inflowing water quality was emphasized.

Keywords: water quality; sensitivity; ELCOM-CAEDYM

---

1. Introduction

The Goyder Water Research Institute project C.1.1 was initiated to fill a gap in the current understanding of the potential impacts of climate change on South Australia. The project seeks to understand climate drivers, downscale global circulation (GCM) model projections of future climate and develop a suite of model applications for the evaluation of climate change impacts on society. Current global circulation model (GCM) projections suggest Australian average temperatures will increase by 1.0 to 5.0 degrees by 2070 (compared to 1980–1999), there will be a decrease in average annual rainfall over southern Australia and there will be an increase in the number of hot days and warm nights [1]. Decreases in winter and autumn wind speed and increases in spring and winter downward solar radiation are also projected, but these projections are subject to large uncertainties [2]. Recent efforts to downscale GCM outputs to the catchment scale have identified the potential for reduced catchment yields as the result of reduced precipitation, changes in rainfall seasonality and increased temperatures [3–5]. Besides issues of water quantity, there are potential impacts of climate change on water quality [6,7]. Reservoirs play a major role in determining the water quality within a given water supply system, as they act as both barriers to (e.g., pathogens) and producers of (e.g., cyanobacteria (toxins, tastes and odors), iron and manganese) water quality hazards [8]. Reservoirs integrate the prevailing hydrology, meteorology, biology and biogeochemistry and the resulting quantity and quality of water is a valuable resource that requires sound management to ensure the utility and sustainability of the source water; water quality models are tools to this end.

The potential impacts of climate change on water quality has been evaluated using integrated modeling schemes which include water quality models [9–13]. Such schemes use a combination of catchment and lake/reservoir models that use meteorological boundary conditions as inputs. The meteorological conditions are altered to represent projected future climate and the resulting simulations are taken to represent the potential impacts of those changed climatic conditions. Too few of these studies have been performed to make generalizations about the potential impacts; both positive and negative influences have been identified. Additionally, the differences in model structure and method make it difficult to compare the different studies directly. There are many sources of uncertainty within such a modeling scheme, including the choice of GCM, emissions scenario, downscaling methodology, and the selection of and rigor of application of the hydrological, constituent and lake/reservoir water quality models, including model structure selection and identification of parameters. Each step in the modeling scheme needs to be thoroughly evaluated to ensure the results can be useful.

It is therefore appropriate to adequately test the response of the proposed reservoir water quality model to changes in the environmental variables expected to change in the future. Formalizing our understanding of the way that water quality variables respond to climate related model inputs is fundamentally important to understanding the outputs we generate from models [13]. As these models will be used to project the impacts of downscaled climate scenarios, it is important that the response of the water quality models to the boundary conditions is understood. Water quality models vary in their data input requirements and often contain options for the sub-model structures they contain, making it difficult to assume that they will be equally sensitive in any given application. Responses of chemical and biological processes to the changes in physical state generated by changes in meteorological inputs are dynamic and interactive and therefore difficult to resolve without resolving individual sensitivities in an explicit analysis.

The outputs from any model are dependent on the inputs. It follows that uncertainty in the inputs, either the boundary conditions or the model parameters, contributes to the uncertainty of the model results. Quantification of the influence of the inputs on the model outputs is known as sensitivity analysis and has been extensively described in the literature. Complex models with many parameters, boundary conditions and long runtimes have particular challenges associated with the analysis of their sensitivity and uncertainty. Consequently a great deal of effort has gone towards developing screening methods to identify sensitive parameters and evaluate their influence on model output [14–17]. Less often the influence of boundary conditions or input data is evaluated. Generally, the error associated with these inputs is considered to be less than the uncertainty associated with model parameters as they are quantities that are generally measured at, or proximal to, the lake or reservoir being modeled, using accurate instrumentation. However the range of meteorological boundary conditions are expected to change in the future [18] and given the non-linear and non-monotonic nature of ecosystem models, their behavior in these conditions is uncertain. As suitable observed validation data cannot exist for unobserved future conditions, model behavior under altered boundary conditions can only be validated against qualitative projected responses of ecosystems. These qualitative responses may be derived from space-for-time approaches, robust ecophysiological conceptual models and response data [19] and ensemble model predictions [20].

Therefore, the goal of this work is to answer the question: Does ELCOM-CAEDYM demonstrate appropriate climatic sensitivity to be used as part of a robust integrated modeling scheme? The responsiveness of the ELCOM-CAEDYM model [21,22] to changes in meteorological boundary conditions was analyzed. A previous application of the model to Happy Valley Reservoir (HVR) was used in conjunction with scenarios with altered environmental forcing of incremental changes in flow, air and water temperature, and wind speed. Responses in water quality variables of primary focus were cyanobacteria and soluble metals; further consideration was given to water temperature and water column stratification due to their important role in determining mixing and the rates of biogeochemical reactions. This work does not constitute a model sensitivity analysis, *sensu stricto*, but evaluates the climatic sensitivity or responsiveness of ELCOM-CAEDYM and compares it to other studies and an empirical climate sensitivity analysis of chlorophyll-*a* in Happy Valley Reservoir.

2. Materials and Methods

2.1. Happy Valley Reservoir

Happy Valley Reservoir (35°04'12" S, 138°34'12" E) is situated to the south of Adelaide, the capital of South Australia (Figure 1). It was created by the construction of an earth wall dam between 1892 and 1897. Following a rehabilitation project from 2002 to 2004, it has a capacity of 11,600 ML, a surface area of 178 hectares and average and maximum depths of 6.5 and 18 m, respectively. It is an off stream reservoir and supplies raw water to South Australia's largest water treatment plant, which produces up to 400 ML of filtered water per day, resulting in a hydraulic retention time of 15–30 days. As HVR is isolated from its natural catchment, it is supplied with water from the Onkaparinga River system via an aqueduct from Clarendon Weir, which is in turn supplied from the much larger Mount Bold Reservoir (35°07'12" S, 138°42'00" E). Mount Bold Reservoir collects water from the Mount Lofty Ranges and is supplemented with water pumped from the River Murray, as are most of South Australia's reservoirs. Happy Valley Reservoir has experienced a range of water quality challenges in the past, with blue-green algae (cyanobacteria) causing taste and odor problems in recent decades. The use of artificial destratification (mixing) and algaecides are used for management in the reservoir, while granular activated carbon used in the water treatment process to reduce taste and odor compound concentrations to acceptable levels in the product water. As HVR is supplied with water from an unprotected catchment (*i.e.*, containing various farming activities and human habitation), vigilance against pathogens is required and loads of nutrients are greater than is generally desirable. During the study period, nutrient concentrations were, total phosphorus, 0.05–0.1 mgL^{-1}; total Kjeldahl nitrogen, 0.5–1.0 mgL^{-1}; filterable reactive phosphorus, 0.005–0.03 mgL^{-1}; ammonia, 0.005–0.05 mgL^{-1} and oxidized nitrogen, 0.05–0.5 mgL^{-1}. The seasonal temperature range is generally between 8–10 °C and 25–27 °C, strong persistent stratification is prevented from occurring by the operation of a bubble plume aerator. Due to the importance of Happy Valley Reservoir to Adelaide's water supply, the South Australian Water Corporation has invested heavily in monitoring and research into the processes influencing water quality.

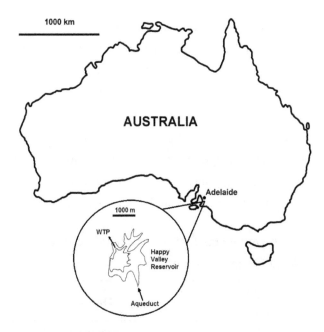

Figure 1. Location of Happy Valley Reservoir. Inset shows 10 m contours of depth and inflow from the aqueduct and the location of the offtake to the water treatment plant (WTP).

2.2. Model Description

The Estuary and Lake Computer Model (ELCOM) is a hydrodynamic model that simulates the temporal behavior of stratified water bodies with environmental forcing. The model solves the unsteady, viscous Navier-Stokes equations for incompressible flow using the hydrostatic assumption for pressure. The simulated processes include baroclinic and barotropic responses, rotational effects, tidal forcing, wind stresses, surface thermal forcing, inflows, outflows, and transport of salt, heat and passive scalars [21]. When coupled with the Computational Aquatic Ecosystem DYnamics Model [22] water quality model, ELCOM can be used to simulate three-dimensional transport and interactions of flow physics, biology and chemistry. ELCOM uses the Euler-Lagrange method for advection of momentum with a conjugate-gradient solution for the free-surface height. Passive and active scalars (*i.e.*, tracers, salinity and temperature) are advected using a conservative ULTIMATE QUICKEST discretization, see [21] and references within for further details.

The Centre for Water Research was previously engaged to apply ELCOM-CAEDYM to Happy Valley Reservoir [23]. Upon delivery, the model was considered appropriate for the simulation of water movement, contaminant transport, algal growth and biogeochemical cycling [23]. ELCOM was applied at three resolutions (25, 50 and 100 m grid sizes); the finest grid to be used for examining short-circuiting and inflow dilution, and the coarser grids for quicker runtimes and running scenarios relating to stratification, algal growth and soluble metal release from sediments (the 100 m grid was used in this study). The hydrodynamic model was validated against temperature sensor data over two periods, 29 June–6 October 2005 and 23 October 2005–8 February 2006. The parameter set for CAEDYM was derived from applications to other Australian reservoirs and some minor calibration of parameters to suit Happy Valley Reservoir. The manual calibration focused on parameters that could not be derived from literature values and included, the density of particulate organic matter, the maximum rate for microbial decomposition of particulate organic phosphorus (nitrogen), the maximum rate of mineralization of dissolved organic phosphorus (nitrogen), the dissolved oxygen ½ saturation constant for nitrification, the rate of denitrification and the phosphorus ½ saturation constant for algal uptake. Some deficiencies in the calibration of the algal growth components of the model remained.

Two algal groups were included in the model structure, representing chlorophytes (green algae) and cyanophytes (blue-green algae). The phytoplankton growth model was parameterized according to literature values, with only a single parameter being manually calibrated for Happy Valley Reservoir (Table 1). Parameters relating to light, temperature, phosphorus uptake and respiratory losses were different between the two phytoplankton groups. All other parameters were shared and derived from literature values. Notably, buoyancy regulation by cyanobacteria was not invoked in the model structure.

Table 1. Phytoplankton group parameters that differentiate the response to ecophysiological drivers in the ELCOM-CAEDYM model set up.

Parameter	Cyanophyte Value	Chlorophyte Value	Description	Reference
μ_{GTH}	0.8	1.2	Maximum growth rate (d^{-1})	[24]
ϑ_{Ag}	1.09	1.07	Temperature multiplier for growth (-)	[25,26]
μ_{RES}	0.09	0.10	Respiration, mortality and excretion (d^{-1})	[27]
K_P	0.009	0.008	P ½ saturation constant ($mg\ L^{-1}$)	Calibrated
I_K	130	100	Light ½ saturation constant ($\mu E\ m^{-2}\ s^{-1}$)	[28]
T_{STD}	24	20	Standard temperature for algal growth (°C)	[29]
T_{OPT}	30	22	Optimum temperature for algal growth (°C)	[29,30]
T_{MAX}	39	35	Maximum temperature for algal growth (°C)	[29]

For this work, the model was not further calibrated or modified beyond the work of Romero *et al.* [23] and therefore no performance metrics are presented. The lack of extensive calibration to HVR water quality dynamics means the results of the study can be considered to be a general test of the response sensitivity of ELCOM-CAEDYM to climate drivers and not an investigation of the likely effects of climate change on water quality in Happy Valley Reservoir.

2.3. Scenarios for Analysis of ELCOM-CAEDYM Climatic Sensitivity

A series of twenty four (24) scenarios were defined, synthetic input data files were generated and ELCOM-CAEDYM simulations were run. As stratification, algal growth and soluble metal concentrations were of key interest, the summer period simulation was used. The 100 m grid version of ELCOM was used to minimize the runtime required, as short-circuiting was not a primary concern of the water quality problems being investigated. The input boundary conditions analyzed were selected to represent the "climate drivers" of precipitation, air temperature and wind speed and are represented by the input files as changes in flow, air and water temperature, and wind speed, respectively (these will be referred to as INFLOW, WIND and TEMP in text). The synthetic input files were generated by applying a linear multiplier, for INFLOW and WIND, and an increment in the case of TEMP (Table 2). Temperature was modified in this fashion to facilitate comparison to potential temperature change magnitudes. For comparison, −5 and +5 degrees correspond to multipliers of 0.8 and 1.25, respectively, at 20 degrees Celsius, similar to the average temperature in the reservoir during the simulations. As ELCOM-CAEDYM will fail if changes to the water budget result in violations in the boundary conditions, changes in the inflow and outflow must be balanced, therefore the outflow (consumption at the offtake) was increased by a corresponding amount. The FLOW scenarios could therefore be considered to represent a change in the consumption of water by the water treatment plant (WTP), rather than changes in precipitation, strictly. This may initially seem artificial; however, as HVR is an offline storage and the inflow to the reservoir is fully regulated by a flume at Clarendon Weir, it can be interpreted as representing changes in demand, especially as a summer period was considered.

Table 2. Boundary condition modifications applied in the sensitivity analysis. A scenario was generated for each change in meteorological variable, resulting in 24 scenarios differing from the base scenario.

Temperature (TEMP) [Increment]	Precipitation (FLOW) [Multiplier]	Wind Speed (WIND) [Multiplier]
−5.0	0.50	0.50
−2.0	0.75	0.75
−1.0	0.90	0.90
−0.5	0.95	0.95
0.5	1.05	1.05
1.0	1.10	1.10
2.0	1.25	1.25
5.0	1.50	1.50

The scenarios were run using the same initial conditions; a "spin-up" period of 1 week was excluded from all summary calculations. As potable water production is the focus of the study, water quality (temperature, suspended solids, chlorophyll, iron and manganese) at the reservoir offtake was analyzed, along with "whole of reservoir" characteristics, such as water temperature and g' (the reduced gravity due to stratification, [21]). Changes in water quality were evaluated as changes in the mean concentration, the maximum concentration and the period of the simulation that the concentration was above a threshold value (green algal and cyanobacterial chlorophyll only, 1 and 10 µg/L, respectively). In order to facilitate the interpretation of the phytoplankton dynamics, summaries of the state variables governing the growth of the two species modeled were calculated as means of the time series values.

2.4. An Empirical Analysis of the Climatic Sensitivity of Chlorophyll-a to Temperature

Historical records of chlorophyll-*a* and water temperature were collated from the primary reservoir surface monitoring location for the period 1998 to 2013. Monthly medians and anomalies were calculated for water temperature and chlorophyll-*a* concentration. The monthly anomalies were normalized to unity, so as to be able to compare directly to modeling results summarized with a similar method. Linear regressions were fitted to the raw anomalies and normalized values, both for the entire year and for the summer months only.

3. Results and Discussion

3.1. Lake Physical Characteristics

The (modeled) physical properties of the lake were altered by the changes in boundary conditions. The degree of stratification, as indicated by average g', was altered in all scenarios; changes in wind speed had a strong negative effect on lake stratification (Table 3). Increasing air and inflowing water temperature resulted in increased reservoir stratification, as did increased flow. Water temperature in the reservoir was not strongly influenced by the INFLOW scenarios, however the WIND and TEMP scenarios had strong effects on the mean of the average, minimum and maximum water temperatures observed over the simulations (Table 3). Only small impacts on reservoir volume and level were observed (not shown).

3.2. Water Quality

An increase in average modeled cyanobacterial chlorophyll was observed with elevated temperature (Figure 2a). The average concentration of reduced soluble iron (FeII) also increased with temperature while soluble manganese was less responsive (Figure 2). Sensitivity responses were close to linear near the origin (±10%), but some became non-linear at the extremes of the scenarios investigated. Exceedance of the threshold selected for cyanobacterial chlorophyll increased approximately linearly with increasing temperature above that of the original scenario, but had little effect below that level (data not shown). The FLOW scenarios had a consistently linear influence on reservoir water quality; increasing average concentrations of chlorophyte and cyanobacterial chlorophyll, MnII and FeII were observed in simulations with reduced flow; only the average concentration of suspended solids

(SSOL1) decreased with decreasing flow (Figure 2b). Changes in maximum modeled values behaved similarly as did duration of exceedance for the chlorophyll variables (not shown).

Table 3. Summary of average physical properties for climatic sensitivity analysis of ELCOM-CAEDYM simulations of Happy Valley Reservoir.

Factor	Increment/ Multiplier	g' (/s²)	Temperature Mean (°C)	Temperature Max (°C)	Temperature Min (°C)
Original	-	0.0502	20.5	21.8	16.5
INFLOW	0.50	0.0481	20.9	22.2	16.6
INFLOW	0.75	0.0490	20.8	22.0	16.6
INFLOW	0.90	0.0496	20.6	21.9	16.5
INFLOW	0.95	0.0498	20.6	21.9	16.5
INFLOW	1.05	0.0503	20.5	21.8	16.5
INFLOW	1.10	0.0505	20.5	21.8	16.6
INFLOW	1.25	0.0510	20.3	21.7	16.6
INFLOW	1.50	0.0513	20.2	21.5	16.6
TEMP	−5.0	0.0454	17.0	18.3	13.4
TEMP	−2.0	0.0481	19.1	20.4	15.9
TEMP	−1.0	0.0490	19.8	21.1	16.2
TEMP	−0.5	0.0495	20.2	21.5	16.4
TEMP	+0.5	0.0505	20.9	22.2	16.7
TEMP	+1.0	0.0511	21.3	22.5	17.0
TEMP	+2.0	0.0524	22.0	23.2	17.3
TEMP	+5.0	0.0571	24.1	25.4	17.5
WIND	0.50	0.0984	22.7	25.9	17.0
WIND	0.75	0.0681	21.5	23.4	17.0
WIND	0.90	0.0560	20.9	22.4	16.7
WIND	0.95	0.0528	20.7	22.1	16.6
WIND	1.05	0.0474	20.4	21.6	16.6
WIND	1.10	0.0452	20.2	21.4	17.2
WIND	1.25	0.0397	19.8	20.8	17.4
WIND	1.50	0.0334	19.3	20.1	17.3

The relationship between WIND and algal growth was obviously non-linear with large increases in the average concentrations of both algal groups with decreasing wind speed (Figure 2c). Cyanobacteria were especially favored by low wind speeds. Reduction of wind speed from 90% to 75% of today's averages resulted in a large increase in the duration of exceedance by cyanobacteria (not shown). The simulated phytoplankton production rates were low (~ 0.1 day^{-1}) compared to what they can potentially be (~ 0.3–0.5 day^{-1}) and probably are in HVR. This was also noted by Romero *et al.* [23]. The simulated whole lake averages of respiration exceeded that of production in cyanobacteria, indicating that they were limited to growing in a limited volume of the lake where sufficient light was available. Elevated temperatures increased cyanobacterial production rates but these increased production rates were kept in check by elevated respiration. There was very little change in the nutrient (N&P) limitation of phytoplankton, even under the INFLOW scenarios; simulated phytoplankton growth was more limited by light availability (Table 4).

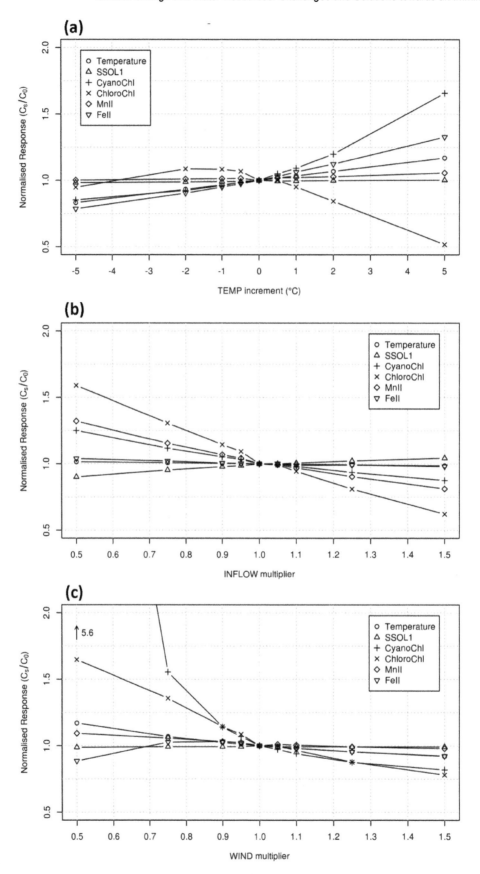

Figure 2. Change in mean modeled water quality values over the summer period in the different sensitivity analysis scenarios where temperature (**a**); rate of inflow and outflow (**b**) or wind speed (**c**) were incrementally changed.

Table 4. Mean cyanobacterial growth characteristics in ELCOM-CAEDYM simulations. The "Limitation by" values indicate the degree of growth limitation by light, phosphorus and nitrogen. It takes a value from 0 to 1; where 1 is unlimited and 0 is completely limited (no growth).

Scenario	Production (day⁻¹)	Respiration (day⁻¹)	Limitation by		
			Light	Phosphorus	Nitrogen
Original	0.080	0.093	0.099	0.915	0.890
INFLOW by 0.5	0.079	0.096	0.095	0.916	0.883
INFLOW by 1.5	0.081	0.091	0.102	0.916	0.890
TEMP by −5	0.061	0.076	0.101	0.917	0.890
TEMP by +5	0.108	0.115	0.106	0.909	0.884
WIND by 0.5	0.083	0.106	0.086	0.923	0.899
WIND by 1.5	0.075	0.087	0.103	0.917	0.889

Production (day⁻¹) shown as (day^{-1}), *Respiration (day⁻¹)* shown as (day^{-1}).

3.3. Implied Model Climatic Sensitivity

These scenarios demonstrate that ELCOM-CAEDYM is responsive to changes in environmental drivers that are expected to change under future climate. The model tested was not heavily calibrated and therefore the results are able to be generalized. The observed sensitivities are consistent with qualitative expectations on the basis of contemporary understanding of reservoir processes; for example, that increased temperature and stratification may; increase the prevalence of cyanobacteria; and result in longer periods of decreased dissolved oxygen concentration and higher dissolved metal concentration. Other authors have observed model climatic sensitivities that resulted in increases in the proportion of cyanobacteria by 1%–7.8% per 1 °C increase in temperature (using the model PROTECH [31]). From a review of the literature of the potential impact of climate on phytoplankton communities, Elliott [13] concluded that projected future climate would result in increased relative abundance of cyanobacteria and changes in the phenology of phytoplankton dynamics but not necessarily an increase in the seasonal amount of phytoplankton biomass. These conclusions are consistent with the responses observed in this study.

Important interactions with nutrient availability exist [32] but this was not investigated here. As an independent factor, nutrient addition (*sensu* INFLOW scenarios) did not have a large effect on the phytoplankton dynamics, presumably because of the lack of nutrient limitation (Table 4). The model tested in this study employed a relatively simple representation of phytoplankton community dynamics; only two main functional groups were represented. Furthermore some physiological mechanisms that facilitate cyanobacterial dominance, despite being available in CAEDYM, were not used in the model application of Romero *et al.* [23]. Greater sensitivity and/or more non-linearity may be expected if these mechanisms (e.g., buoyancy regulation) were implemented.

The environmental drivers that were manipulated in the scenarios were not investigated factorially, however they are not completely independent; changes in mean and maximum water temperature occurred in the INFLOW and WIND scenarios (Table 3). This complicates the interpretation of model outputs without extensive comparison of individual simulations; an effort not warranted by the goals of this study. The scenarios were arbitrarily selected to quickly develop a picture of the sensitivity of the model to changed boundary conditions. As such, the important environmental drivers of dilution and nutrient loading are confounded in the multiplication of inflow volumes. Inflow scenarios assumed the same constituent concentrations and therefore the higher flow scenarios had higher nutrient loads. However as chlorophyll concentrations decreased as flow increased; it is apparent that dilution was a more important driver of algal biomass than nutrient load and availability. Despite this, the prediction that phytoplankton growth is rarely limited by nutrient availability may suggest that reducing the external load may be an option for reducing algal growth. The internal load was not investigated as part of this study but given the short water retention time of the reservoir, it is probably of minor importance, compared to the external load. The reduction of nutrient availability represents a potential strategy for adaptation to climate change and the likely negative effects on water quality resulting from increased cyanobacterial growth. Water quality models, such as ELCOM-CAEDYM, have an important role to play in determining the potential benefit of a nutrient reduction program.

3.4. Empirical Reservoir Climatic Sensitivity

Linear regression between water temperature and chlorophyll median monthly anomalies did not resolve slope estimates significantly different from zero (0.105 ± 0.134, $Pr(>|t|) = 0.43$). The weak positive slope estimate combined with a poor predictive relationship ($R^2 = 0.0142$) demonstrates that surface water temperature did not play an important role in determining total chlorophyll in this period (Figure 3b); it also demonstrates that total chlorophyll was not negatively correlated with water temperature, as implied by the water quality model (Figure 3a). This might suggest that deficiencies in definition of model structure or parameter identification have resulted in a non-behavioral model response (one not consistent with our expectations). These deficiencies could, for example, be found in the parameterization of the temperature response functions for growth, or be the product of the over-simplification of the phytoplankton community. This remains speculative, as this simple comparison cannot resolve the differences between the processes structuring algal growth in the model scenarios as compared to those operating over a longer period and in different years, within the reservoir. It must further be noted that the empirical analysis is limited to (monthly) anomalies less than +2 °C and so could not explore the full range of (annual) anomalies as defined by the model scenarios.

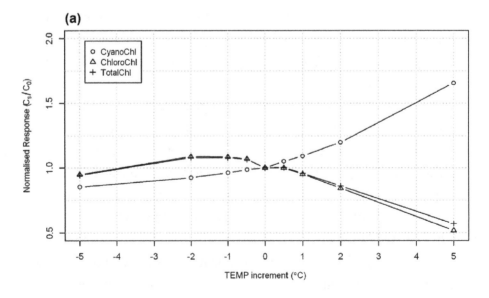

Figure 3. *Cont.*

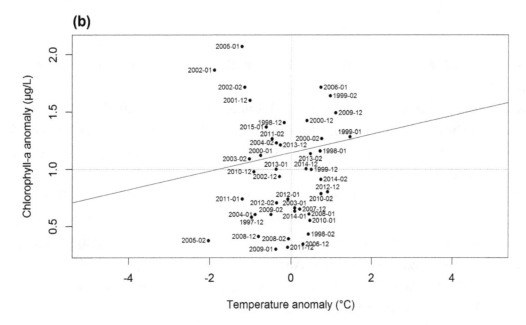

Figure 3. Comparison of (**a**) model derived climate sensitivity to (**b**) empirical reservoir climate sensitivity of chlorophyll-*a* to temperature in summer (December, January, February). In panel (**b**) each point represents the unity normalized anomaly from the monthly median value calculated over the period 1998–2013 and is labeled as yyyy-mm.

4. Conclusions

This study demonstrated that ELCOM-CAEDYM is sensitive to climate drivers and suitable for use in climate impact studies. Rigorous evaluation of the impact of selection of model structures and parameter values on the conclusions drawn from scenarios conducted with altered boundary conditions is advised. This study highlighted factors likely to be important in determining phytoplankton growth in Happy Valley Reservoir. Further it demonstrates that the water quality of the source waters will be of major importance to the reservoir water quality dynamics.

Acknowledgments

This work was funded by the Goyder Water Research Institute Project C1.1. Development of an agreed set of climate projections for South Australia.

Author Contributions

Leon van der Linden conceived and implemented the model climatic sensitivity analysis, the empirical climate sensitivity analysis, analysed the data, interpreted the results and prepared the manuscript; Robert Daly prepared the original input data for ELCOM-CAEDYM and contributed to interpreting the results and preparing the manuscript; Mike Burch conceived the work program, secured funding, managed the research group, and contributed to preparing the manuscript.

References

1. CSIRO. *State of the Climate 2012*; CSIRO and Bureau of Meteorology: Canberra, Australia, 2012; p. 12.

2. CSIRO. *Climate Change in Australia*; CSIRO: Canberra, Australia, 2007.

3. Heneker, T.; Cresswell, D. *Potential Impact on Water Resource Availability in the Mount Lofty Ranges due to Climate Change*; Science, Monitoring and Information Division, Department for Water: Adelaide, Australia, 2010.

4. Charles, S.P.; Heneker, T.; Bates, B.C. Stochastically downscaled rainfall projections and modelled hydrological response for the Mount Lofty Ranges, South Australia. In Proceedings of Water Down Under 2008, Adelaide, Australia, 15–18 April 2008; Lambert, M., Daniell, T., Leonard, M., Eds.; Engineers Australia: Adelaide, Australia, 2008; pp. 428–438.

5. Green, G.; Gibbs, M.; Wood, C. *Impacts of Climate Change on Water Resources, Phase 3: Northern and Yorke Natural Resources Management Region DFW Technical Report 2011/03*; Government of South Australia, Department for Water: Adelaide, Australia, 2011; Volume 1.

6. Whitehead, P.G.; Wilby, R.L.; Battarbee, R.W.; Kernan, M.; Wade, A.J. A review of the potential impacts of climate change on surface water quality. *Hydrol. Sci. J.* **2009**, *54*, 101–123.

7. Delpla, I.; Jung, A.-V.; Baures, E.; Clement, M.; Thomas, O. Impacts of climate change on surface water quality in relation to drinking water production. *Environ. Int.* **2009**, *35*, 1225–1233.

8. Brookes, J.; Burch, M.; Hipsey, M.; Linden, L.; Antenucci, J.; Steffensen, D.; Hobson, P.; Thorne, O.; Lewis, D.; Rinck-Pfeiffer, S.; *et al. A Practical Guide to Reservoir Management*; Cooperative Research Centre for Water Quality and Treatment: Adelaide, Australia, 2008; p. 116.

9. Arheimer, B.; Andréasson, J.; Fogelberg, S.; Johnsson, H.; Pers, C.B.; Persson, K. Climate change impact on water quality: Model results from southern Sweden. *Ambio* **2005**, *34*, 559–566.

10. Mimikou, M.A.; Baltas, E.; Varanou, E.; Pantazis, K. Regional impacts of climate change on water resources quantity and quality indicators. *J. Hydrol.* **2000**, *234*, 95–109.

11. Saloranta, T.; Forsius, M.; Järvinen, M.; Arvola, L. Impacts of projected climate change on the thermodynamics of a shallow and a deep lake in Finland: Model simulations and Bayesian uncertainty analysis. *Hydrol. Res.* **2009**, *40*, 234–248.

12. Thorne, O.M.; Fenner, R.A. Modelling the impacts of climate change on a water treatment plant in South Australia. *Water Sci. Technol. Water Supply* **2008**, *8*, 305–312.

13. Elliott, J.A. Is the future blue-green? A review of the current model predictions of how climate change could affect pelagic freshwater cyanobacteria. *Water Res.* **2012**, *46*, 1364–1371.

14. Arhonditsis, G.B.; Perhar, G.; Zhang, W.; Massos, E.; Shi, M.; Das, A. Addressing equifinality and uncertainty in eutrophication models. *Water Resour. Res.* **2008**, *44*, doi:10.1029/2007WR005862.

15. Makler-Pick, V.; Gal, G.; Gorfine, M.; Hipsey, M.R.; Carmel, Y. Sensitivity analysis for complex ecological models—A new approach. *Environ. Model. Softw.* **2011**, *26*, 124–134.

16. Saltelli, A. Making best use of model evaluations to compute sensitivity indices. *Comput. Phys. Commun.* **2002**, *145*, 280–297.

17. Campolongo, F.; Cariboni, J.; Saltelli, A. An effective screening design for sensitivity analysis of large models. *Environ. Model. Softw.* **2007**, *22*, 1509–1518.

18. Schlabing, D.; Frassl, M.A.; Eder, M.M.; Rinke, K.; Bárdossy, A. Use of a weather generator for simulating climate change effects on ecosystems: A case study on Lake Constance. *Environ. Model. Softw.* **2014**, *61*, 326–338.

19. Paerl, H.W.; Paul, V.J. Climate change: Links to global expansion of harmful cyanobacteria. *Water Res.* **2012**, *46*, 1349–1363.

20. Trolle, D.; Hamilton, D.P.; Pilditch, C.A.; Duggan, I.C.; Jeppesen, E. Predicting the effects of climate change on trophic status of three morphologically varying lakes: Implications for lake restoration and management. *Environ. Model. Softw.* **2011**, *26*, 354–370.

21. Hodges, B.; Dallimore, C. *Estuary, Lake and Coastal Ocean Model: ELCOM v2.2 User Manual*; Centre for Water Research, University of Western Australia: Crawley, Australia, 2007.

22. Hipsey, M.R.; Romero, J.R.; Antenucci, J.P.; Hamilton, D.P. *Computational Aquatic Ecosystem Dynamics Model CAEDYM v2.3 User Manual*; Centre for Water Research, University of Western Australia: Crawley, Australia, 2006.

23. Romero, J.; Antenucci, J.; Okley, P. *Happy Valley Reservoir Modelling Study—Final Report*; Centre for Water Research, University of Western Australia: Crawley, Australia, 2005; p. 41.

24. USCE. *CE-QUAL-R1: A Numerical One-Dimensional Model of Reservoir Water Quality; User's Manual*; Instruction Report E-82-1 (Revised Edition); Department of the Army, U.S. Corps Engineers: Washington, DC, USA, 1995; p. 427.

25. Krüger, G.H.J.; Eloff, J.N. The influence of light intensity on the growth of different Microcystis isolates. *J. Limnol. Soc. South. Afr.* **2010**, *3*, 21–25.

26. Coles, J.F.; Jones, R.C. Effect of temperature on photosynthesis-light response and growth of four phytoplankton species isolated from a tidal freshwater river. *J. Phycol.* **2000**, *36*, 7–16.

27. Schladow, S.; Hamilton, D. Prediction of water quality in lakes and reservoirs: Part II—Model calibration, sensitivity analysis and application. *Ecol. Modell.* **1997**, *96*, 111–123.

28. Hamilton, D.P.; Schladow, S.G. Prediction of water quality in lakes and reservoirs. Part I—Model description. *Ecol. Model.* **1997**, *96*, 91–110.

29. Griffin, S.L.; Herzfeld, M.; Hamilton, D.P. Modelling the impact of zooplankton grazing on phytoplankton biomass during a dinoflagellate bloom in the Swan River Estuary, Western Australia. *Ecol. Eng.* **2001**, *16*, 373–394.

30. Robarts, R.D.; Zohary, T. Temperature effects on photosynthetic capacity, respiration, and growth-rates of bloom-forming cyanobacteria. *N. Z. J. Mar. Freshw. Res.* **1987**, *21*, 391–399.

31. Elliott, J.A.; Jones, I.D.; Thackeray, S.J. Testing the sensitivity of phytoplankton communities to changes in water temperature and nutrient load, in a temperate lake. *Hydrobiologia* **2006**, *559*, 401–411.

32. Mooij, W.M.; Janse, J.H.; Senerpont Domis, L.N.; Hülsmann, S.; Ibelings, B.W. Predicting the effect of climate change on temperate shallow lakes with the ecosystem model PCLake. *Hydrobiologia* **2007**, *584*, 443–454.

12

Climatic Characteristics of Reference Evapotranspiration in the Hai River Basin and their Attribution

Lingling Zhao [1,2]**, Jun Xia** [2,3,*]**, Leszek Sobkowiak** [4] **and Zongli Li** [5]

[1] Guangzhou Institute of Geography, No.100 Xianliezhong Road, Guangzhou 510070, China
    E-Mail: linglingzhao@foxmail.com

[2] Key Laboratory of Water Cycle & Related Land Surface Processes, Institute of Geographic
    Sciences and Natural Resources Research, Chinese Academy of Sciences, Beijing 100101, China;

[3] State Key Laboratory of Water Resources & Hydropower Engineering Science, Wuhan University
    Wuhan 430072, China

[4] Institute of Physical Geography and Environmental Planning, Adam Mickiewicz University,
    Poznan 61-680, Poland; E-Mail: lesob@amu.edu.pl

[5] General Institute of Water Resources and Hydropower Planning and Design, Ministry of Water
    Resources of China, Beijing 100120, China. E-Mail: lizongli@igsnrr.ac.cn

*   Author to whom correspondence should be addressed; E-Mail: xiaj@igsnrr.ac.cn;

Abstract: Based on the meteorological data from 46 stations in the Hai River Basin (HRB) from 1961–2010, the annual and seasonal variation of reference evapotranspiration was analyzed. The sensitivity coefficients combined with the detrend method were used to discuss the dominant factor affecting the reference evapotranspiration (ET_o). The obtained results indicate that the annual reference evapotranspiration is dominated by the decreasing trends at the confidence level of 95% in the southern and eastern parts of the HRB. The sensitivity order of climatic variables to ET_o from strong to weak is: relativity humidity, temperature, shortwave radiation and wind speed, respectively. However, comprehensively considering the sensitivity and its variation strength, the detrend analysis indicates that the decreasing trends of ET_o in eastern and southern HRB may be caused mainly by the decreasing wind speed and shortwave radiation. As for the relationship between human activity and the trend of ET_o, we found that ET_o decreased more significantly on the plains than in the mountains. By contrast, the population density increased more considerably from 2000 to 2010 on the plains than in the mountains. Therefore, in this paper, the correlation of the spatial variation pattern between ET_o and population was further analyzed. The spatial correlation coefficient between population and the trend of ET_o is -0.132, while the spatial correlation coefficient between the trend of ET_o and elevation, temperature, shortwave radiation and wind speed is 0.667, 0.668, 0.749 and 0.416, respectively. This suggests that human activity has a certain influence on the spatial variation of ET_o, while natural factors play a decisive role in the spatial variation of reference evapotranspiration in this area.

Keywords: climatic variation; reference evapotranspiration; Penman–Monteith method; Hai River Basin

1. Introduction

Hydrologists have found that climate change has resulted in some changes in the water cycle [1–3]. One major challenge of recent hydrological modeling activities is the assessment of the effects of climate change on the terrestrial water cycle [4]. Hydrological models are usually based on the calculation of reference evapotranspiration and reducing it to the actual evapotranspiration by considering the soil moisture status [5] or the number of days since the last rainfall event [6]. Therefore, analyzing how climate change affects reference evapotranspiration (ET_o) is critical for understanding the impact of climate change on the hydrological cycle. According to Allen et al. [7], ET_o is the evapotranspiration from the reference surface, which is a hypothetical grass reference crop with an assumed crop height of 0.12 m, a fixed surface resistance of 70 s m^{-1} and an albedo of 0.23.

The Hai River Basin (HRB) is one of seven largest river basins and also one of the most developed areas in China, with the population accounting for about 10% of the nation's total. The middle and lower reaches of the basin are important wheat production regions in China. This region has a semi-humid and semi-arid climate and has been strongly influenced by human activities. The annual precipitation is 539 mm, while the annual pan evaporation is 1100 mm, making the basin vulnerable to climatic variations [7]. In recent decades, several eco-environmental problems in that area have come to the fore under the combined impacts of climate change and intensified human activities. Water resources in the HRB are currently used for irrigation, aquaculture and industries. Due to the very limited available water resources in the basin, water has been diverted from other basins to supply it for agriculture and to maintain essential ecosystem functions [8].

In order to understand how climatic variables affect ET_o, some studies have been carried out to evaluate evapotranspiration in the context of climate change. Zheng et al. [9] analyzed the cause of the decreased pan evaporation during 1957–2001 in the HRB, and found the reason to be the declining wind speed. Xu et al. [10] proved that the decreasing wind speed and net radiation were responsible for the ET_o changes in the Changjiang River Basin of China. Liu [11], who analyzed the pan evaporation from 1955 to 2000 in China, found that the decrease in solar radiance was most likely the driving force of the reduced pan evaporation in China. Furthermore, sensitivity analysis has also been performed on the impacts of climate change [12–18]. However, the temporal pattern of ET_o is not only influenced by the sensitivity of climatic variables, but also by their variation patterns.

The spatial pattern of ET_0 in HRB has not been addressed in the literature, yet. In this study, we calculated ET_0 using the FAO-56 Penman–Monteith equation and analyzed the temporal-spatial pattern in ET_0 and its driving variables. Attribution analysis was then performed to quantify the contribution of each input variable to ET_0 variation. The objective of this paper is to exhibit the temporal-spatial variation pattern of ET_0 over the past 50 years in the HRB, then to detect the reason for these characteristics and to quantify the contribution of the climatic variation to ET_0.

2. Study Area and Data

The Hai River Basin is located in north China and is one of seven largest river basins in the country. The basin is bounded in the north by Mount Tangshan, in the west by Mount Taihang and in the east by the Bohai Sea. Land surface elevation in the mountainous north and west of the study area is generally above 2000 m a.s.l. On the floodplains, however, surface elevation hardly exceeds 100 m a.s.l. The basin occupies an area of 3.2×10^5 km^2 (34.9–42.8° N, 112.0–119.8° E) and includes five provinces and the two megacities of Beijing and Tianjin (Figure 1). Climatically, the HRB belongs to the East Asian monsoon region. The annual mean temperature varies from 8 °C to 12 °C, while annual precipitation is about 539 mm; relative humidity varies from 50% to 70%.

Figure 1. The location of the Hai River Basin (HRB).

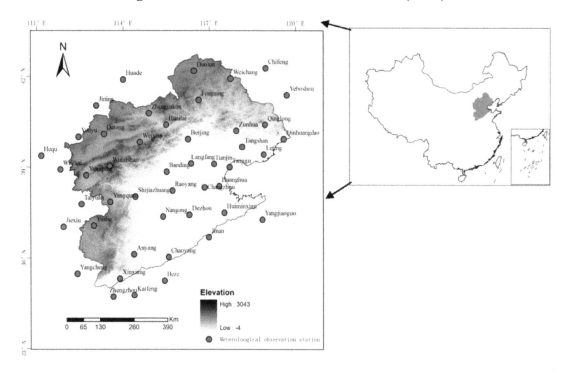

Data from 46 National Meteorological Observatory stations included daily observations of the maximum, minimum and mean air temperatures (T_{max}, T_{min}, T_a), wind speed (U), relative humidity (Rh) and sunshine duration (n) for the period from 1960 to 2010 and pan evaporation (E_{pan}) for 1960–2001. E_{pan} was measured using a metal pan, 20 cm in diameter and 10 cm high, installed 70 cm above the ground. The data have been provided by the National Climatic Center of China Meteorological Administration. The locations of the stations are shown in Figure 1, while the details of the stations are listed in Table 1. In Table 1, the annual ET_o is the average value from 1960 to 2010 calculated using the FAO-56 Penman–Monteith method.

Table 1. Basic data on the investigated stations in HRB.

Name	No.	Longitude (°)	Latitude (°)	Elevation (m)	Name	No.	Longitude (°)	Latitude (°)	Elevation (m)
Wutaishan	53588	113.53	39.03	2896	Changzhou	54616	116.83	38.33	10
Weixian	53593	114.57	39.83	910	Tanggu	54623	117.72	39.00	3
Yuanping	53673	112.72	38.73	828	Huanghua	54624	117.35	38.37	7
Shijiazhuang	53698	114.42	38.03	81	Nangong	54705	115.38	37.37	27
Yangquan	53782	113.55	37.85	742	Dezhou	54714	116.32	37.43	21
Yushe	53787	112.98	37.07	1041	Huiminxian	54725	117.53	37.50	12
Anyang	53898	114.37	36.12	76	Chaoyang	54808	115.58	36.03	43
Xinxiang	53986	113.88	35.32	73	Huade	53391	114.00	41.90	1483
Duolun	54208	116.47	42.18	1245	Shiyu	53478	112.45	40.00	1346
Fengning	54308	116.63	41.22	660	Jiying	53480	113.07	41.03	1419
Weichang	54311	117.75	41.93	843	Hequ	53564	111.15	39.38	862
Zhuangjiakou	54401	114.88	40.78	724	Wuzhai	53663	111.82	38.92	1401
Huailai	54405	115.50	40.40	537	Taiyuan	53772	112.55	37.78	778
Zunhua	54429	117.95	40.20	55	Jiexiu	53863	111.92	37.03	744
Qinglong	54436	118.95	40.40	227	Yangcheng	53975	112.40	35.48	660
Qinhuangdao	54449	119.60	39.93	2	Chifeng	54218	118.97	42.27	568
Beijing	54511	116.28	39.93	54	Yeboshou	54326	119.70	41.38	662
Langfang	54518	116.38	39.12	9	Yangjiaogou	54736	118.85	37.27	6
Tianjin	54527	117.17	39.10	3	Jinan	54823	116.98	36.68	52
Tangshan	54534	118.15	39.67	28	Heze	54906	115.43	35.25	50
Leting	54539	118.90	39.42	11	Zhengzhou	57083	113.65	34.72	110
Baoding	54602	115.52	38.85	17	Kaifeng	57091	114.38	34.77	73
Raoyang	54606	115.73	38.23	19	Datong	53487	113.33	40.10	1067

3. Methodologies

3.1. Penman–Monteith Method

The Penman–Monteith method recommended by FAO (Food and Agriculture Organization) [19] as the standard method for determining reference evapotranspiration was used in this study. The method was selected because it is physically based and explicitly incorporates both physiological and aerodynamic parameters.

$$ETo = \frac{0.408\Delta\left(R_n - G\right) + \gamma \frac{900}{\left(T+273\right)} U_2 (e_s - e_a)}{\Delta + \gamma(1 + 0.34 U_2)} \tag{1}$$

where, ET_o is the reference evapotranspiration (mm/day); R_n is net radiation at the crop surface (MJ/m^2/day); G is soil heat flux density (MJ/m^2/day); T is mean daily air temperature (°C), U_2 is wind speed at 2 m height (m/s); e_s is saturation vapor pressure (kPa); ($e_s - e_a$) is the saturation vapor pressure deficit (kPa); Δ is the slope of vapor pressure (kPa/°C) and γ is the psychometric constant (kPa/°C).

The computation of all data required for the calculation and relevant procedures are given in Chapter 3 of the FAO Paper 56 [19].

3.2. Trend Detection and Sensitivity Analysis Method

The rank-based nonparametric Mann–Kendall statistical test [20,21] is commonly used for trend detection, because of its robustness for non-normally distributed and censored data, which are frequently encountered in hydroclimatic time series. In this method, the test statistic, Z, is as follows:

$$Z = S / \sigma_s^2 \tag{2}$$

$$\text{with } S = \sum_{i=1}^{n-1} \sum_{j=i+1}^{n} \text{sgn}(x_j - x_i) \tag{3}$$

$$\sigma_s^2 = \frac{n(n-1)(2n+5) - \sum_{i-1}^{n} e_i i(i-1)(2i+5)}{18} \tag{4}$$

$$\text{sgn}(x) = \begin{cases} 1 & \text{if } x > 0 \\ 0 & \text{if } x = 0 \\ -1 & \text{if } x < 0 \end{cases} \tag{5}$$

Equation (2) gives the standard deviation of S with correction for ties in the data, with e_i denoting the number of ties of extent i. The upward or downward trend in the data is statistically significant if $|Z| > \mu_{1-\alpha/2}$, where $\mu_{1-\alpha/2}$ is the $(1-\alpha/2)$ quantity of the standard normal distribution and when $\alpha = 0.05, u_{1-\alpha/2} = 1.96$. Positive Z indicates an increasing trend in the time series, while negative Z, a decreasing one.

Original measurements of air temperature (T_a), wind speed (U) and relative humidity (Rh) were chosen for the sensitivity analyses. The fourth applied variable is shortwave radiation (R_s). This is because shortwave radiation is one of the input variables in a number of semi-physical and semi-empirical equations that are used to derive the net energy flux required by the Penman method [22]. Following the procedure described by Allen [19], R_s can be estimated with the following formula that relates surface shortwave radiation to extraterrestrial radiation and daily sunshine duration:

$$R_s = (a_s + b_s \frac{n}{N}) R_a \tag{6}$$

where R_s is shortwave radiation, n is daily sunshine duration (h); N is maximum possible duration of sunshine or daylight hours (h); n/N is relative sunshine duration; R_a is extraterrestrial radiation and a_s and b_s are the regression constants. The recommended values a_s = 0.25 and b_s = 0.75 were used in this study.

In multivariate models, different variables have different dimensions and different ranges of values, which make it difficult to compare the sensitivity by partial derivatives. Consequently, the partial derivative is transformed into a non-dimensional form:

$$Sx = \lim_{\Delta x/x} \left(\frac{\dfrac{\Delta ETo}{ETo}}{\dfrac{\Delta x}{x}} \right) = \frac{\partial ETo}{\partial x} \cdot \frac{x}{ETo} \tag{7}$$

Basically, a positive/negative sensitivity coefficient of a variable indicates that ET_o increases/decreases as the variable increases; the larger the sensitivity coefficient, the larger the effect a given variable has on ET_o.

3.3. Spatial Correlation Coefficient

Correlation coefficients depict the spatial relationship between two datasets. The correlation between two variables is a measure of dependency between these variables. It is the ratio of the covariance between the two datasets divided by the product of their standard deviations. Because it is a ratio, it is a unit-less number. The equation to calculate the correlation is [23]:

$$Corr_{ij} = \frac{Cov_{ij}}{\delta_i \delta_j} \tag{8}$$

where, Cov_{ij} is the covariance; δ_i, δ_j are the standard deviations of dataset i and j, respectively.

The calculated covariance matrix in this paper contains values of variances and covariances. The variance is a statistical measure showing how much variance there is from the mean. The remaining entries within the covariance matrix are the covariances between all pairs of the input datasets. The following formula is used to determine the covariance between datasets i and j:

$$Cov_{ij} = \frac{\sum\limits_{k=1}^{N} (Z_{ik} - \mu_i)(Z_{jk} - \mu_j)}{N-1} \tag{9}$$

where, Z_{ik}, Z_{jk} are the values of dataset i and dataset j, respectively in location k; μ_i, μ_j are the average values of datasets i and j, respectively; i, j is the order of dataset; N is the number of dataset; K denotes a particular location.

Correlation ranges from $+1$ to -1. A positive correlation indicates a direct relationship between two datasets, such as when the cell values of one datasets increase, the cell values of another datasets are also likely to increase. A negative correlation means that one variable changes inversely to the other. A correlation of zero means that two datasets are independent of one another.

3.4. Detrend Method

The variation pattern of ETo is determined by multi-climatic variables, including their sensitivity to ETo and variation fluctuations. The detrend method is a combination method that considers both the sensitivity coefficient and the fluctuation of the climatic variables.

The detrend method is a way of quantifying the contribution of climatic variables to the annual variation of ET_o. This method shows the contribution in graphs and vividly describes how the climatic variables influence ET_o. In this study, the following steps were performed: (1) use of the simple linear regression method to detect the changing slope of the main climatic variables; (2) detection of the significance of the slope by the t-test; (3) removal of the significant slope of main climatic variables to make them stationary time series; (4) recalculation of reference evapotranspiration using each time the original series of three variables and the detrend data of one variable; (5) comparison of the results with the original reference evapotranspiration; the observed difference is considered as the influence of those variables on the trend [22].

4. Results and Discussion

4.1. Correlation between ET_o and E_{pan}

Figure 2 shows the monthly correlation coefficients (R^2) between ET_o and E_{pan} in HRB. As can be seen, the lowest correlation coefficient is 0.93, while the highest is 0.97. Spatially, mountain areas in the northwest and coastal areas in the southeast have higher correlation coefficients than plains in the central part of the study area. Mountain and coastal areas are more humid than plains, which is consistent with the research conclusion of Brutsaert [24] that in humid areas, reference evaporation has a better relationship with pan evaporation.

Figure 2. Correlation between ET_o and E_{pan} in the HRB.

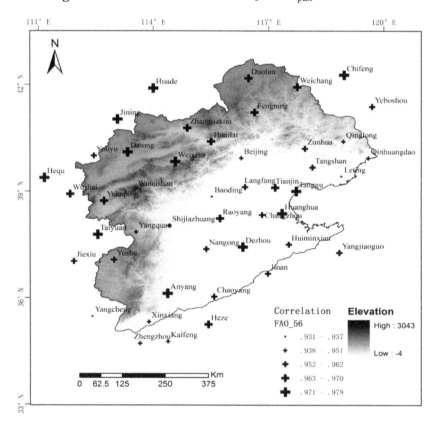

4.2. Spatial-Temporal Variation of ET_o in HRB

To study the spatial distribution of the trend of ET_o from 1960 to 2010 in the HRB, the Mann–Kendall test was used for each station to establish the ET_o trends. The trends of annual and seasonal ET_o were tested at the 95% confidence level. Decreasing trends in annual ET_o were observed at 28 stations located mostly on the eastern and southern plains of the HRB. However, increasing trends in the annual ET_o were observed at three sites (Datong, Wutaishan, Weixian) located in the western mountain region of the HRB (Figure 3).

Figure 3. The annual trend of ET_o in the HRB.

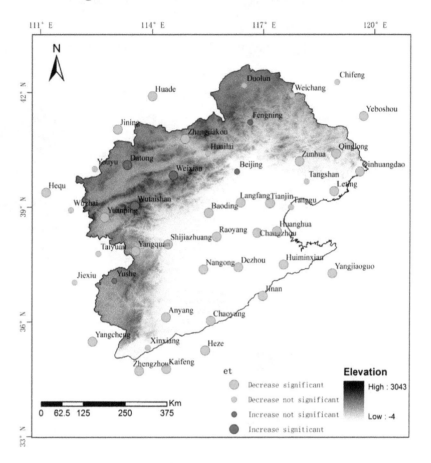

As for the variation pattern of seasonal ET_o, Figure 4 shows large differences in the spatial distribution of trends in seasonal ET_o in the HRB. As can be seen, stations with decreasing trends in spring are mainly distributed in the southeastern plain region of the study area. The significantly increasing trend of ET_o was found only at Wutaishan station in the western mountain region of the HRB. Twenty-nine stations characterized by decreasing trends of ET_o in summer are concentrated mainly in the eastern and southern regions of the HRB. Only one station is dominated by the increasing trend of ET_o in summer. In autumn, only nine out of 46 stations show a trend in ET_o. Among them, eight stations present decreasing trends, while one station shows an increasing trend. The changing trend patterns in winter are similar to those in autumn: 16 out of 46 stations are dominated by the decreasing winter ET_o, while two stations in the western plateau region display the increasing trend.

4.3. Variation Pattern of Climatic Variables

Shortwave radiation decreases in the whole basin, and most trends are significant at the 0.05 significance level (Figure 5). The maximum and minimum temperatures increased in the whole HRB (Figure 5). The maximum temperatures in the southern mountain area increase more significantly than on the northern plains; the average p-value of the Mann–Kendall test for the whole basin is 2.9

(Table 2). The minimum temperatures increase more obviously than the maximum ones; the average p-value of the Mann–Kendall test for the whole basin is five. Relative humidity decreases in most of the basin, and this trend is significant in most sites of the mountain areas. As to the wind speed, the decreasing trends are significant in most parts of the basin at the 0.05 significance level.

Figure 4. The seasonal trend of ET_o in the HRB. (**a**) Spring; (**b**) Summer; (**c**) Winter; (**d**) Autumn.

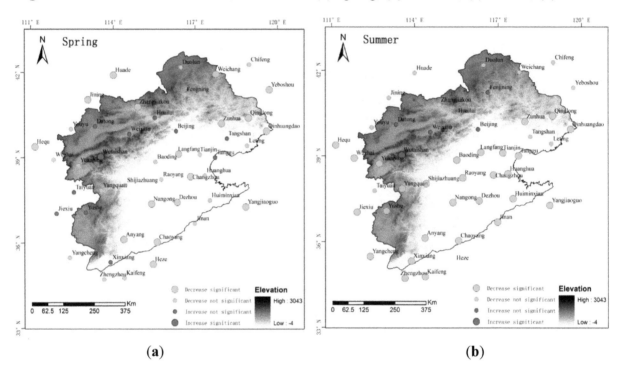

(**a**)   (**b**)

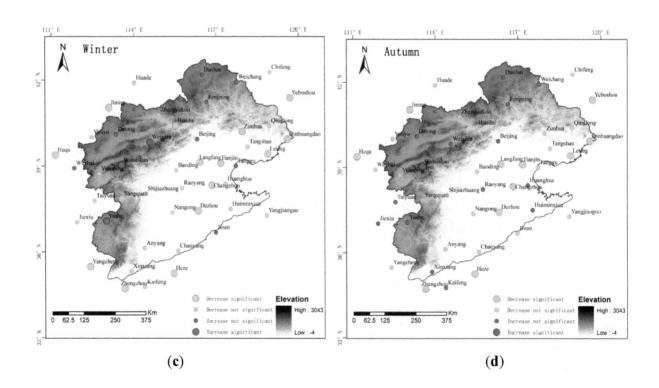

(**c**)   (**d**)

Figure 5. The trend of main climatic variables in the HRB. (**a**) The trend of T_a; (**b**) The trend of R_s; (**c**) The trend of U; (**d**) The trend of Rh.

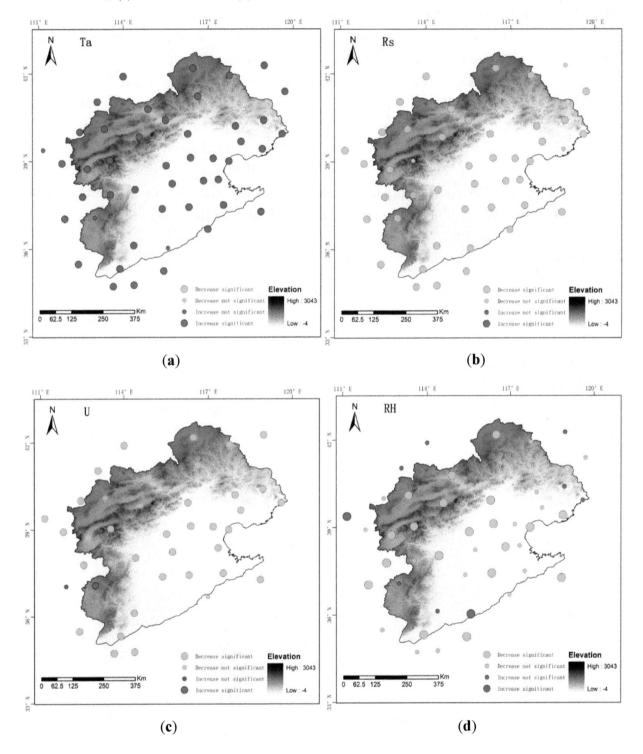

(**a**)

(**b**)

(**c**)

(**d**)

Table 2. The Mann–Kendall test value of the main climate variables at the three stations.

Station name	T_a	U	Rh	R_s
Datong	4.78	−1.55	−1.81	−2.54
Wutaishan	4.30	−5.21	−3.57	−0.54
Weixian	4.87	−0.92	−3.07	−2.04
Mean value of the HRB	4.16	−4.12	−1.41	−3.94

The increasing temperature and decreasing relative humidity will make ET_o increase, while the decreasing short wave radiation and wind speed will make ET_o decrease. Therefore, the decreasing trends of ET_o in the HRB suggest that the increasing temperature and decreasing relative humidity slightly discount the decreasing trend, but do not change its direction; the decreasing shortwave radiation and wind speed commonly result in the decreasing ET_o in the southeastern coastal area of the HRB.

In order to detect the reasons for the increasing trend of ET_o in Datong, Wutaishan and Weixian stations, this paper analyzed the climate variables at these three locations. We found a Mann–Kendall test value of temperature as large as at the other investigated stations. However, relative humidity decreased faster, while shortwave radiation and wind speed (except Wutaishan) were slower compared to the mean value calculated for the whole HRB. Therefore, it can be concluded that the increasing temperature, decreasing faster relative humidity and decreasing slower shortwave radiation and wind speed result in the increasing trend of ET_o at these three stations.

4.4. Sensitivity of Climatic Variables

Since ET_o is an important indicator of climatic changes, the sensitivity coefficient was used in this study to analyze how climatic variables affect ET_o. The non-dimensional form of the sensitivity coefficient was employed to estimate the sensitivity of climatic variables in the HRB. Figure 6 gives the annual sensitivity coefficient of four climatic variables (T_a, Rh, R_s and U) to ET_o from 1960 to 2010 estimated by the FAO-56 Penman–Monteith method. Figure 6 suggests that temperature, wind speed and short wave radiation are less sensitive in the mountain areas than on the coastal plains, while relative humidity is less sensitive in the area closest to the coast than on the plains and in the mountains.

Figure 6. The sensitivity coefficients of climatic variables in the HRB. (**a**) The sensitivity coefficients of T_a; (**b**) The sensitivity coefficients of Rh; (**c**) The sensitivity coefficients of U; (**d**) The sensitivity coefficients of R_s.

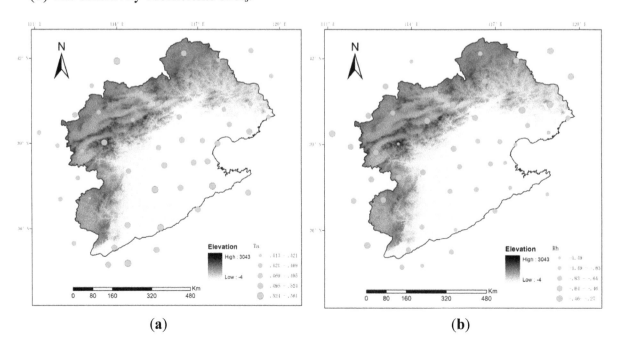

(**a**)   (**b**)

Figure 6. *Cont.*

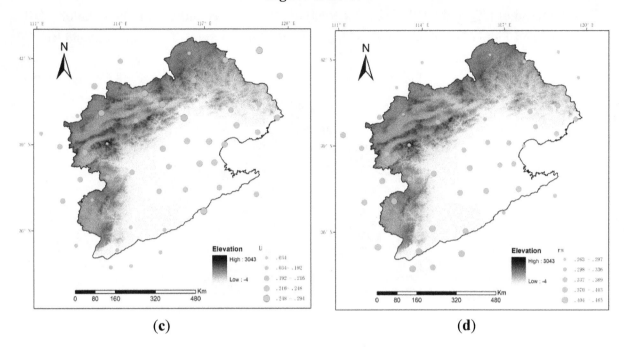

(c)   (d)

As for the trend of sensitivity to the climatic variables, the decreasing trends are found in temperature, short wave radiation and relative humidity, while the increasing trend is in wind speed. As variations in sensitivity coefficients to main climatic variables are detected before and after 1990, so the data series of sensitivity coefficients were divided into two groups of 1960–1990 and 1991–2010, respectively, to calculate mean sensitivity coefficients to climatic variables. The obtained results show that before 1990, the sensitivity coefficients to T_a, Rh, U and R_s were 0.500, −0.641, 0.205 and −0.357, respectively, while after 1990, they were 0.477, −0.586, 0.221 and 0.349, respectively.

4.5. ET_o with Detrend Climatic Variables

Figure 7 gives the trend of ET_o estimated with the detrend data series. The trends of ET_o with detrend temperature are decreasing, except one station, while the variation pattern of ET_o with detrend relative humidity is similar to the original ET_o. This shows that there are larger differences between the original ET_o and the recalculated one with detrend wind speed or detrend shortwave radiation than that with the detrend temperature or relative humidity. This suggests that the decreasing wind speed and shortwave radiation may be the main causes of the decreasing ET_o in the HRB. As for the decreasing trend of shortwave radiation, previous studies have shown that the decrease in global radiation is the most likely the cause, which is a regional phenomenon. By examining the regional total radiation in eastern China, Zhang [25] concluded that the regional total radiation is decreasing due to the increased air pollution in that area. Another study by Liu [26] also proved that air pollution may result in the decrease of R_s in HRB. Therefore, they speculate that aerosols may play a critical role in the decrease of solar radiation in China.

Figure 7. The annual trend of ET_o with detrend climatic variables in the HRB. (**a**) The annual trend of ET_o in the HRB; (**b**) The trend of ET_o with detrend U; (**c**) The trend of ET_o with detrend T_a; (**d**) The trend of ET_o with detrend R_s; (**e**) The trend of ET_o with detrend Rh.

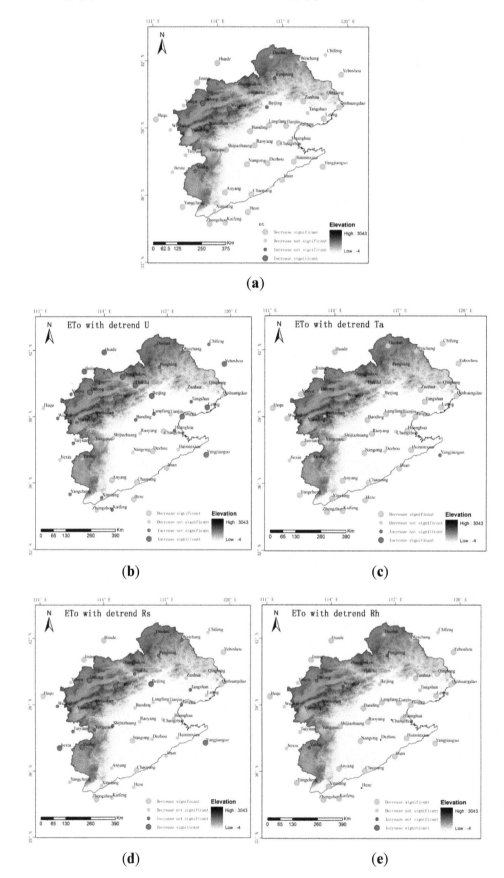

4.6. Relationship between ET_o and Human Activity

The relationship between the variation pattern of ET_o and human activity refers to how human activity affects climatic variables. In order to detect the relationship between human activity and the variation of climatic variables, the investigated sites were divided into two groups; elevation was taken as the sole criterion of that division. Areas with an elevation higher than 500 m a.s.l. are usually defined as mountains, while below 500 m a.s.l. as plains. In general, the population density is higher and human activity is more intense on the plains than in the mountain areas. Figure 8 gives the GDP (Gross Domestic Product) increase from 2000 to 2010 and the distribution of plains in the HRB, respectively. As can be seen, the GDP increased noticeably from 2000 to 2010 on the plains of the HRB. This may suggest that human activity is concentrated on the plains. At the same time, the decrease of ET_o is more visible on the plains than in the mountains.

Figure 8. The GDP variation and distribution of plain areas in the HRB. (**a**) GDP variation during 2010 to 2000 in HRB; (**b**) DEM below 500 m a.s.l. in HRB.

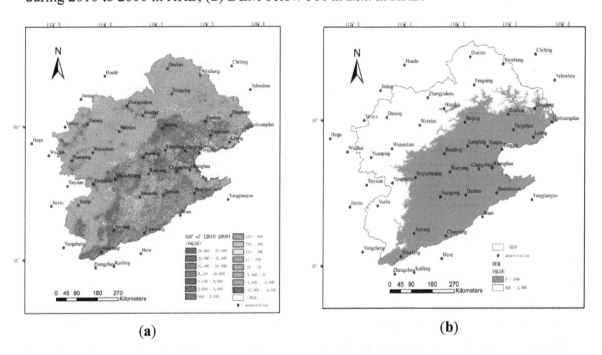

(**a**)   (**b**)

Table 2 gives the trend of the investigated climatic variables in these two elevation groups. It can be seen that the trend of the shortwave radiation, wind speed and minimum temperature on the plains is more obvious than that in the mountain areas. The detrend results show that the decreasing shortwave radiation and wind speed are the main causes of the decreasing ET_o on the plains of the HRB. Therefore, we deduce that there is a relationship between the human activity and the decreasing shortwave radiation and wind speed. Liu *et al.* [26] found that during the period from 1957 to 2008, the solar radiation decreased significantly in the HRB and that the trend was more significant in the densely populated areas than in the sparsely populated ones. The spatial distribution of the aerosol index increase is consistent with the solar radiation decrease. The aerosol increase resulting from human activities was an important reason for the decrease in solar radiation. This phenomenon was also found by some other studies [27–29]. As for the wind speed, Jiang *et al.* [30] analyzed the change of wind speed over China from 1956 to 2004. They found that the effect of urbanization on annual MWS (Mean Wind Speed) is more and more obvious with time, especially in the 1980s. It corresponds with the substantial development of urbanization. At the same time, the declining trend of mean wind speed at the urban stations is more serious than that at the rural ones over the last 50 years. This is coincident with the results given in Table 3.

Table 3. The trend of climatic variables in different elevation groups.

Elevation	R_s	T_{max}	T_{min}	Rh	U
>500 m a.s.l.	−3.689	3.698	4.727	−1.157	−3.773
<500 m a.s.l.	−4.601	2.170	5.198	−1.348	−4.739

Furthermore, in order to determine the relationship between ET_o and human activities, this paper analyzed with the use of the spatial correlation coefficient method the spatial correlation between ET_o, population and meteorological factors, which have a definite effect on ET_o. The results show that the spatial correlation coefficient between population and ET_o is −0.132, while the spatial correlation coefficients between population, R_s and Rh are −0.307 and −0.144, respectively. Moreover, the spatial correlation coefficients between elevation and ET_o is 0.667, while the spatial correlation coefficients between ET_o, R_s, U and T_a are 0.749, 0.416 and 0.668, respectively. This proves that human activity has a certain influence on the spatial variation of ET_o, while natural factors play a decisive role in the spatial variation character of reference evapotranspiration (Table 4).

Table 4. The spatial correlation coefficients between ET_o and different factors.

R_s	U	T_a	Rh	Population	Elevation
0.749	0.416	0.668	−0.267	−0.132	0.667

With increasing altitude and decreasing pressure, the atmosphere is getting thinner and the heat radiation losses become faster. Therefore, the temperature decreases with increasing altitude. At the same time, when the sunlight scattering and reflection decrease, the solar radiation increases. While the proportion of water vapor in the air is relatively large, it is usually concentrated in the lower parts of the atmosphere. Relatively small air moisture content at high altitudes results in low relative humidity there. Friction caused by the uneven Earth surface retards the air flow, so that the wind speed is relatively smaller near the ground. With higher elevation (up to 6000 m a.s.l.), the wind speed becomes less impacted by the uneven ground.

5. Conclusions

In this paper, the spatial and temporal characteristics of annual and seasonal ET_o (1960–2010) in the Hai River Basin (HRB) were examined, and the possible causes of changes in ET_o were detected. The following conclusions may be drawn from the study:

(1) Most stations in the HRB have decreasing trends in the annual ET_o at a confidence level of 95%. These stations are distributed mainly in the southern and eastern coastal areas of HRB. Three stations (Datong, Wutaishan and Weixian) in the western area of HRB show significant increasing trends in the annual ET_o. As for the seasonal changes, similar characteristics with

respect to the annual ET_0 were identified only in summer, while during the other three seasons (spring, autumn and winter), the trends were less obvious.

(2) The spatial patterns of the Mann–Kendall trends of the annual meteorological variables show that the maximum and minimum temperatures increase significantly at the 0.05 significance level. However, the increase of the minimum temperature is more apparent than that of the maximum ones all over the basin. Wind speed and shortwave radiation show decreasing trends in the whole basin, and the trends are significant in the eastern and southern parts of the HRB. The sensitivity analysis shows that relativity humidity is the most sensitive variable to ET_0, followed by temperature, shortwave radiation and wind speed as the least sensitive to ET_0 in the whole HRB.

(3) Comprehensively considering the sensitivity and variation strength of the meteorological variables, the detrend analysis indicates the decreasing trends in ET_0 dominant in the eastern and southern area of HRB. These may be caused mainly by the behavior of wind speed and shortwave radiation. Meanwhile, the obtained detrend results suggest that the increasing temperature is the main cause of the increasing trend of ET_0 in Datong, Wutaishan and Weixian stations.

(4) The spatial correlation coefficient between population and the trend of ET_0 is -0.132, and the correlation coefficient between the trend of ET_0 and natural factors is even higher. This suggests that human activity has a certain influence on the spatial variation of ET_0, while natural factors play a decisive role in the spatial variation character of reference evapotranspiration in this area.

Acknowledgments

The authors are grateful to the National Climate Center of China for providing climatic data. Creative Talents Fund of Guangzhou Institute of Geography, Young Talents Fund of Guangdong Academy of Sciences (rcjj201303) and the Natural Science Foundation of China (No. 51279140) contributed to this research.

Author Contributions

Ling-ling Zhao is the first person to complete the paper, is responsible for the calculation of the paper, written and modified. Leszek Sobkowiak responsible for checking the language and spelling, Jun Xia and Zongli Li give recommendations in the paper forming process.

References

1. Okechukwu, A.; Luc, D.; Souley, Y.K.; le, B.E.; Ibrahim, M.; Abdou, A.; Théo, V.; Bader, J.; Moussa, I.B.; Gautier, E.; *et al.* Increasing river flows in the Sahel? *Water* **2010**, *2*, 170–199.

2. Huntington, T.G. Evidence for intensification of the global water cycle: Review and synthesis. *J. Hydrol.* **2006**, *319*, 83–95.

3. Changchun, X.; Yaning, C.; Yuhui, Y.; Hao, X.M.; Shen, Y.P. Hydrology and water resources variation and its response to regional climate change in Xinjiang. *J. Geogr. Sci.* **2010**, *20*, 599–612.

4. Zhao, L.L.; Xia, J.; Xu, C.Y.; Wang, Z.G.; Sobkowiak, L.; Long, C. Evapotranspiration estimated methods in hydrological simulation. *J. Geogr. Sci.* **2013**, *23*, 359–369.

5. Feddes, R.A.; Kowalik, P.J.; Zaradny, H. *Simulation of Field Water Use and Crop Yield*; Centre for Agricultural Publishing and Documentation: Wageningen, Netherlands 1978.

6. Ritchie, J.T. Model for predicting evaporation from a row crop with incomplete cover. *Water Resourc. Res.* **1972**, *8*, 1204–1213.

7. Bo, T.; Ling, T.; Shaozhong, K.; Lu, Z. Impacts of climate variability on reference evapotranspiration over 58 years in the Haihe river basin of North China. *Agric. Water Manag.* **2011**, *98*, 1660–1670.

8. Jun, X.; Huali, F.; Chesheng, Z.; Cunwen, N. Determination of a reasonable percentage for ecological water-use in the Haihe River Basin, China. *Pedosphere* **2006**, *16*, 33–42.

9. Zheng, H.X.; Liu, X.M.; Liu, C.M.; Dai, X.Q.; Zhu, R.R. Assessing contributions to panevaporation trends in Haihe River Basin, China. *J. Geophys. Res. Atmos.* **2009**, *114*, doi:10.1029/2009JD012203.

10. Xu, C.Y.; Gong, L.B.; Jiang, T.; Chen, D.L. Decreasing reference evapotranspiration in a warming climatea case of Changjiang (Yangtze) River catchment during 1970–2000. *Adv. Atmos. Sci.* **2006**, *23*, 513–520.

11. Liu, B.H.; Xu, M.; Henderson, M.; Gong, W.G. A spatial analysis of pan evaporation trends in China, 1955–2000. *J. Geophys. Res.* **2004**, *109*, doi: 10.1029/2004JD004511.

12. Bormann, H. Sensitivity analysis of 18 different potential evapotranspiration models to observed climatic change at German climate stations. *Clim. Chang.* **2011**, *104*, 729–753.

13. Coleman, G..; DeCoursey, D.G. Sensitivity and model variance analysis applied to some evaporation and evapotranspiration models. *Water Resourc. Res.* **1976**, *12*, 873–879.

14. Gao, G.; Chen, D.L.; Ren, G.Y.; Chen, Y.; Liao, Y.M. Spatial and temporal variations and controlling factors of potential evapotranspiration in China: 1956–2000. *J. Geogr. Sci.* **2006**, *16*, 3–12.

15. Gong, L.B.; Xu, C.Y.; Chen, D.L.; Halldin, S.; Chen, Y.Q. David. Sensitivity of the Penman–Monteith reference evapotranspiration to key climatic variables in the Changjiang (Yangtze River) basin. *J. Hydrol.* **2006**, *329*, 620–629.

16. Goyal, R.K. Sensitivity of evapotranspiration to global warming: A case study of arid zone of Rajasthan (India). *Agric. Water Manag.* **2004**, *69*, 1–11.

17. Hupet, F.; Vanclooster, M. Effect of the sampling frequency of meteorological variables on the estimation of the reference evapotranspiration. *J. Hydrol.* **2001**, *243*, 192–204.

18. McCuen, R.H. A sensitivity and error analysis of procedures used for estimating evaporation1. *JAWRA J. Am. Water Resourc. Assoc.* **1974**, *10*, 486–497.

19. Allen, R.G.; Pereira, L.S.; Raes, D.; Smith, M. *Crop Evapotranspiration-Guidelines for Computing Crop Water Requirements-FAO Irrigation and Drainage Paper 56*; Food and Agriculture Organization: Rome, Italy, 1998; 1–15.

20. Kendall, M.G. *Rank Correlation Measures*; Charles Griffin: London, UK, 1975; Volume 202.

21. Mann, H.B. Nonparametric tests against trend. *Econ. J. Econ. Soc.* **1945**, *13*, 245–259.

22. Xu, C.Y.; Gong, L.B.; Jiang, T.; Chen, D.L.; Singh, V.P. Analysis of spatial distribution and temporal trend of reference evapotranspiration and pan evaporation in Changjiang (Yangtze River) catchment. *J. Hydrol.* **2006**, *327*, 81–93.

23. Chong, S.Z. *The Probabilistic Approach in Water Science and Technology*; Science Press: Beijing, China, 2010. (In Chinese)

24. Brutsaert, W. Land-surface water vapor and sensible heat flux: Spatial variability, homogeneity, and measurement scales. *Water Resourc. Res.* **1998**, *34*, 2433–2442.

25. Zhang, Y.Q.; Liu, C.M.; Tang, Y.H.; Yang, Y.H. Trends in pan evaporation and reference and actual evapotranspiration across the Tibetan Plateau. *J. Geophys. Res. Atmos.* **2007**, *112*, doi: 10.1029/2006JD008161.

26. Liu, C.M.; Liu, X.M.; Zheng, H.X.; Zeng, Y. Change of the solar radiation and its causes in the Haihe River Basin and surrounding areas. *J. Geogr. Sci.* **2010**, *20*, 569–580.

27. Grimenes, A.A.; Thue-Hansen, V. The reduction of global radiation in south-eastern Norway during the last 50 years. *Theor. Appl. Climatol.* **2006**, *85*, 37–40.

28. Qian, Y.; Kaiser, D.P.; Leung, L.R.; Xu, M. More frequent cloud-free sky and less surface solar radiation in China from 1955 to 2000. *Geophys. Res. Lett.* **2006**, *33*, doi: 10.1029/2005GL024586.

29. Qian, Y.; Wang, W.G.; Leung, L.R.; Kaiser, D.P. Variability of solar radiation under cloud-free skies in China: The role of aerosols. *Geophys. Res. Lett.* **2007**, *34*, doi: 10.1029/2006GL028800.

30. Jiang, Y.; Luo, Y.; Zhao, Z.C.; Tao, S.W. Changes in wind speed over China during 1956–2004. *Theor. Appl. Climatol.* **2010**, *99*, 421–430.

Permissions

List of Contributors

Sisira S. Withanachchi and Angelika Ploeger
Department of Organic Food Quality and Food Culture, University of Kassel, Nordbahnhofstr. 1a, 37213 Witzenhausen, Germany

Sören Köpke
Institute for Social Sciences, Technische Universität Braunschweig, Bienroder Weg 97, D-38106 Braunschweig, Germany

Chandana R. Withanachchi
Department of Archaeology and Heritage Management, Rajarata University, Mihintale 50300, Sri Lanka

Ruwan Pathiranage
Eco-collective Research Association, Colombo 00200, Sri Lanka

Hong-Ming Liu
Department of Civil and Disaster Prevention Engineering, National United University, Miaoli 36003, Taiwan

Wen-Cheng Liu
Department of Civil and Disaster Prevention Engineering, National United University, Miaoli 36003, Taiwan
Taiwan Typhoon and Flood Research Institute, National Applied Research Laboratories, Taipei 10093, Taiwan

Alec Zuo, Sarah Ann Wheeler and Henning Bjornlund
School of Commerce, University of South Australia, Adelaide 5000, Australia

Robert Brooks
Faculty of Business and Economics, Monash University, Caulfield East 3145; Australia

Edwyna Harris
Department of Economics, Monash University, Clayton 3168, Australia

Mohammed Saif Al-Kalbani and Martin F. Price
Centre for Mountain Studies, Perth College, University of the Highlands and Islands, Crieff Road, Perth PH1 2NX, UK

Asma Abahussain
Department of Natural Resources and Environment, College of Graduate Studies, Arabian Gulf University, Manama 26671, Bahrain

Mushtaque Ahmed
Department of Soils, Water & Agricultural Engineering, College of Agricultural and Marine Sciences, Sultan Qaboos University, Al-Khod 123, Oman

Timothy O'Higgins
Scottish Association for Marine Sciences, University of the Highlands and Islands, Scottish Marine Institute, Dunstaffnage, Argyll PA37 1QA, UK

Tianhong Li and Yuan Gao
College of Environmental Sciences and Engineering, Peking University, Beijing 100871, China
Key Laboratory of Water and Sediment Sciences, Ministry of Education, Beijing 100871, China

Rupak Aryal and Simon Beecham
Centre for Water Management and Reuse, School of Natural and Built Environments, University of South Australia, Mawson Lakes, SA 5095, Australia

Alistair Grinham
School of Civil Engineering, The University of Queensland, St Lucia, QLD 4072, Australia

Ken Okamoto
United Graduate School of Agricultural Sciences, Kagoshima University, 1-21-24 Korimoto, Kagoshima-shi, Kagoshima 890-0065, Japan

Kazuhito Sakai, Tamotsu Nakandakari and Shinya Nakamura
Faculty of Agriculture, University of the Ryukyus, 1 Senbaru, Nishihara-cho, Okinawa 903-0213, Japan

Hiroyuki Cho
Faculty of Agriculture, Saga University, 1 Honjo-machi, Saga 840-8502, Japan

Shota Ootani
Eight-Japan Engineering Consultants Inc.; 33-11 Honcho 5 Chome, Nakano-ku, Tokyo 164-0012, Japan

Eihab Fathelrahman
Department of Agribusiness and Consumer Sciences, College of Food and Agriculture, United Arab Emirates University, Al-Magam Campus, Al-Ain, United Arab Emirates

Amalia Davies and Stephen Davies
Pakistan Strategy Support Program (PSSP), the International Food Policy Research Institute (IFPRI), IFPRI-PSSP Office #006, Islamabad, Pakistan

James Pritchett
Department of Agricultural and Resource Economics, Colorado State University, Campus Mail 1172, Fort Collins, CO 80523, USA

Mary Akurut
Department of Civil Engineering, KU Leuven, Kasteelpark Arenberg 40, bus 2448, Leuven 3001, Belgium
Department of Civil and Environmental Engineering, Makerere University Kampala, Kampala 00256, Uganda

Charles B. Niwagaba
Department of Civil and Environmental Engineering, Makerere University Kampala, Kampala 00256, Uganda

Patrick Willems
Department of Civil Engineering, KU Leuven, Kasteelpark Arenberg 40, bus 2448, Leuven 3001, Belgium
Department of Hydrology and Hydraulic Engineering, Vrije Universiteit Brussel, Pleinlaan 2, Brussels 1050, Belgium

Shaowei Ning, Hiroshi Ishidaira and Parmeshwar Udmale
International Research Center for River Basin Environment (ICRE), University of Yamanashi, Takeda 4-3-11, Kofu, Yamanashi 400-8511, Japan

Yutaka Ichikawa
Department of Civil and Earth Resources Engineering, Kyoto University, C1, Kyoto-Daigaku-Katsura, Nishikyo-ku, Kyoto-shi, Kyoto 615-8540, Japan

Leon van der Linden, Robert I. Daly and Mike D. Burch
South Australian Water Corporation, Adelaide, SA 5000, Australia

Lingling Zhao
Guangzhou Institute of Geography, No.100 Xianliezhong Road, Guangzhou 510070, China
Key Laboratory of Water Cycle & Related Land Surface Processes, Institute of Geographic Sciences and Natural Resources Research, Chinese Academy of Sciences, Beijing 100101, China

Jun Xia
Key Laboratory of Water Cycle & Related Land Surface Processes, Institute of Geographic Sciences and Natural Resources Research, Chinese Academy of Sciences, Beijing 100101, China
State Key Laboratory of Water Resources & Hydropower Engineering Science, Wuhan University Wuhan 430072, China

Leszek Sobkowiak
Institute of Physical Geography and Environmental Planning, Adam Mickiewicz University, Poznan 61-680, Poland

Zongli Li
General Institute of Water Resources and Hydropower Planning and Design, Ministry of Water Resources of China, Beijing 100120, China

Index